Consumer's Guide
to a Brave New World

Consumer's Guide
to a Brave New World

Wesley J. Smith

ENCOUNTER BOOKS
SAN FRANCISCO

First edition published in 2004 by Encounter Books, an activity of Encounter for Culture and Education, Inc., a nonprofit, tax exempt corporation.

Encounter Books website address: www.encounterbooks.com

Manufactured in the United States and printed on acid-free paper.

The paper used in this publication meets the minimum requirements of ANSI/NISO Z39.48-1992 (R 1997)(Permanence of Paper).

Library of Congress Cataloging-in-Publication Data

Smith, Wesley J.
    Consumer's guide to a brave new world / by Wesley J. Smith
        p.   cm.
        ISBN 1-893554-99-6 (alk. paper)
        1. Biotechnology—Social aspects.   2. Genetic engineering—Social
    aspects.   I. Title.
    TP248.23 .S657 2004
    306.4'61—dc22

                                                    2004056273

10 9 8 7 6 5 4 3 2

To Mark Pickup—

"Blessed are the pure in heart

for they shall see God."

Contents

. . . .

Introduction

· ·

THIS BOOK IS ABOUT THE HUMAN FUTURE. More precisely, it is about the power of biotechnology to affect the human future by harnessing our bodies at the cellular level, first to employ them as raw material in the manufacture of medicines, and ultimately to transform what we commonly regard as human nature. In these pages we will explore the scientific, moral and business aspects of cloning, embryonic and adult stem cell research and genetic engineering; we will consider the potential for miracle medical cures and the prospect of fabricating "designer babies"; and we will ponder the tremendous impact these activities are bound to have on us all.

Few contemporary issues are as important, or as controversial. The decisions we make today about biotechnology will materially impact each of our lives and those of our children; this potent new science will transform society, and conceivably even what it means to be human. For better and for worse—perhaps very much worse—biotechnology is developing the godlike power not only to improve our health and extend our longevity, but also to remake the biology of the human species.

Issues that were once the exclusive domain of science fiction writers are now everyday headlines: In early 2003, the Raelians, a science cult fixated on flying saucers, made international headlines by claiming it had created the first cloned human babies;[1] biotech companies' press releases boast about developing medical products manufactured from tissues derived from the destruction of nascent human life;[2] and with the successful mapping of the human genome, some bioethicists and futurists now urge that we use the information we have gained to manufacture offspring enhanced for greater intelligence, strength and beauty. No wonder some observers fear—while others eagerly antici-pate—that genetic engineering will steer us toward the creation of a "posthuman" race.[3]

Experiments involving the very building blocks of life are already being conducted. A biotech company has patented a herd of goats genet-ically engineered to include spider DNA, so the herd's ewes produce milk containing spider-web silk; the feat has been repeated with cows and hamsters.[4] Cloning researchers claimed to have produced an early embryo that was mostly human, but also part animal.[5] A mouse exists today with millions of human cells in its brain.[6] Some biotechnologists and bioethicists look forward to the time when similar research will be conducted on human beings.

But such surreal experimentation is far from being the whole story. Biotechnology holds extraordinary promise for alleviating human suf-fering in a greatly improved and fully human future. Scientists are learn-ing how to treat illnesses utilizing the patients' own cells as healing medicines: in early human trials, patients' tissues have been put to work treating damaged hearts, Parkinson's disease and multiple sclerosis.[7]

Still, many of these emerging technologies may have "dual use" potentials, leading many to worry that they will be harnessed in the serv-ice of malevolent purposes. Self-described "transhumanists" advocate that we seize control of human evolution by designing genetic enhance-ments to "improve" the human race. As an example of this sort of think-ing, the Nobel Prize-winning scientist James D. Watson, co-discoverer of the DNA double helix, has claimed that genetically enhanced people will someday "dominate the world."[8] Are there echoes here of the long-discredited "master race" idea?

With biotechnology offering so much good and threatening such harm, the question comes down to this: Does society have the capacity and the will to guide and regulate science so as to keep it on a path that not only improves our lives, but also proceeds with deep respect for the inherent dignity and intrinsic value of our humanity? Few contempo-rary issues are so consequential. As Leon Kass, chairman of the Presi-dent's Council on Bioethics, said to me:

> All of the natural boundaries are up for grabs. All of the boundaries that have defined us as human beings, boundaries between a human being

and an animal on one side, and between a human being and a superhuman being—or a god—on the other. The boundaries of life, the boundaries of death. The normal human relations that are founded upon ties born of sexual reproduction, so that every newborn child is somehow the fusion of two lines going back to time immemorial. These are the questions of the twenty-first century—and nothing could be more important.[9]

Translating the Scientific Jargon

Even people unacquainted with the scientific details sense that we have suddenly entered unexplored territory, and they are filled with unease. Yet many are hesitant to participate in the debate over how, if at all, biotechnology should be regulated; and if so, how to mark out the ethical frontier beyond which its activities should not be allowed to stray. Their reluctance is entirely understandable: the issues seem horrendously complex, the science so abstruse that most people think they can never gain sufficient understanding to enable them to reach informed opinions.

But in fact, though it may take years of training and vast expertise before scientists can carry out the astonishing experiments we will be discussing, an understanding of the moral issues involved is accessible to almost everyone; once the special language so often used by biotechnologists to describe their work is translated into simpler words, the fog lifts and matters become much clearer.

Unraveling the science is just the first step, however. To get an adequate grasp of the latest advances and come to an informed opinion about their moral consequences, we'll have to scrutinize the powerful financial forces that are pushing us toward accepting human cloning, embryonic stem cell research, genetic engineering and the like. In other words, we must follow the money.

Undoubtedly, many of the best scientists in the vanguard of research are motivated by curiosity, enthusiasm for discovery and altruism; likewise the entrepreneurs, private venture capitalists and government administrators who underwrite their work. But idealistic emotions rarely go unmixed, and some—perhaps most—researchers not only hope, but expect, to find a pot of gold at the end of the biotech rainbow; indeed, the almost universal human desire to get rich may be the strongest driving force behind the dizzying pace of today's advances.

There's nothing wrong with prospering from one's work, of course: business enterprise has brought many and vast benefits to humankind. Unfortunately, in this debate as in so many others past and present, our pocketbooks may also influence our morality—and not necessarily for the better. In this connection, it's worth recalling the nineteenth-century defenses of the "morality" of slavery, or (more recently) the amoral apologetics of tobacco companies.

The real—or imagined—financial rewards that spur on biotech research are a story too seldom told; particularly so when industry participants are frequently presented as objective analysts, unselfishly devoted to the pursuit of knowledge and the curing of disease. In contrast, the skeptics' arguments are derided as being driven by religion, or at least by an outmoded conservative philosophy.

This picture, subtly and not-so-subtly touted by the media, is highly misleading. As I will show, some of the most committed biotech advocates are (in the nice phrase of *National Journal* reporter Neil Munro) as much "academic entrepreneurs" as they are objective scientists, and their resistance to reasonable regulation of biotechnology may be strongly influenced by personal financial considerations.[10] The yearning for fame and fortune on the one hand, and a fervent belief in virtually unfettered scientific inquiry on the other, interlock like pieces of a jigsaw puzzle.

The same and more can be said about some of the more radical biotech companies that insist on pursuing the most controversial areas of research. Starved for investment capital, and hoping to lure investors to write big checks to finance their work (or better yet, to get taxpayers to foot the bill), some have resorted to crass public-relations campaigns—hence the fawning stories by reporters who not only exaggerate accomplishments but weave the false impression that new medical treatments for some of the worst diseases may be just over the horizon. Such deceitful promises cruelly stoke the hopes of very ill people and their families; and they unfairly distort, in my view, the political debate that will determine the impact that biotech will have on the human future.

The Ideology of Science

Ideology also factors into biotech controversies. While science writers and media editorialists are eager to claim, for example, that religious fervor and/or political conservatism primarily motivate critics of human cloning (let's ignore for a moment the fact that many opponents of these futuristic technologies are committed secularists and/or political liberals), what often goes unreported is the strong, endemic bias of the pro-research lobby: a pervasive "scientism."

In the newest definition of the term, "scientism" is a philosophy or a belief system in which scientific inquiry is seen as much more than a dispassionate way of gaining knowledge about the natural world and discovering physical laws.* Rather, unfettered science is seen as *the* fount of earthly salvation. Or, to put it another way, the means and manner of our pursuit of knowledge are removed from any moral context, and the quest itself—a morally neutral undertaking—becomes its own purpose and meaning. Moreover, just as religious beliefs do in fact motivate some critics of the biotech exploitation of human life, an equally passionate secular fervor impels some its supporters. Seeing the universe through a materialist prism, disdaining faith of any kind except their own, loudly proclaiming that religious values have no place in debates over public policy, these enthusiasts embrace science as the ultimate liberator of humankind from "superstition in all its forms, and especially in the form of religious belief."[11]

This quasi worship of—rather than respect for—science can easily corrupt its proper pursuit. The fact is that in recent years science has, to a disturbing degree, become political. Although some blame this development on those who have raised alarums about potential abuses of biotech, the latter are far from being solely responsible. In growing measure, scientists themselves have become fervent policy advocates, disguised as impartial investigators. As the science journal *Nature* recently warned:

---

*According to Michael Shermer, editor-in-chief of *Skeptic* magazine, "Scientism is a scientific worldview that encompasses natural explanations for all phenomena, eschews supernatural and paranormal explanations, and embraces empiricism and reason as the twin pillars of a philosophy of life appropriate for an Age of Science." ("The Shamans of Scientism," *Scientific American*, June 2002.)

As political battles are waged through "science," many scientists are willing to adopt tactics of demagoguery and character assassination as well as, or even instead of, reasoned argument. Science is becoming yet another playing field for power politics, complete with the trappings of media spin and a win-at-all-costs attitude. Sadly, much of what science can offer policy-makers, and hence society, is being lost.[12]

In researching this book, I have come to believe that the debate over stem cells, human cloning, genetic engineering and other controversial avenues of research is, at its core, about far more than biotechnology. I believe that we are in the midst of an epochal struggle whose outcome will determine our moral priorities. At stake is whether a Promethean science for science's sake will trump all other considerations, including long-established beliefs about right and wrong; or whether the scientific enterprise will serve us with humility, and with the understanding that morality can never be jettisoned—most particularly when one is manipulating the building blocks of human life.

The Stakes Are High

There are some biotechnologists who promise a miraculous future, a golden age of extended lives and enhanced capacities—if only we free ourselves from constraining notions of right and wrong, which are said to be rooted in outdated superstitious hang-ups. The disabled will walk, diabetics will throw away their insulin, those with Parkinson's disease will regain control over their bodies. We may come near to achieving the age-old dream of immortality. Or we may evolve into superbeings of heightened intelligence, astounding athleticism and sublime beauty.

But others, myself included, see need for great caution. On the one hand, no one objects to researching the use of adult tissues and stem cells. Such research may reap most of the medical benefits sought by advocates of human cloning, with none of the moral cost. Indeed, these uncontroversial technologies are already ameliorating human suffering and curing human diseases. For example, transplanted adult eye stem cells have restored sight to patients blinded by chemical burns or certain rare diseases.[13] On the other hand, embryonic stem cells—which have as yet produced lesser results than adult stem cells—pose serious moral questions. How will it affect our future if human embryos and perhaps fetuses are regarded as mere natural resources to be exploited and harvested like a corn crop or prize cattle herd? Is it right and ethical to create human life with the intention of destroying it, even if the overarching purpose is to obtain knowledge and treat human illness? Or, as President George W. Bush put it so well in his 2003 State of the Union address, should we not be wary of beginning and ending human lives "as an experiment"? And what would be the consequences of permitting genetic enhancements? Would it really lead to an improved humanity—or, as some fear, to a "new eugenics" that would create hierarchies of perceived human worth, where the "enhanced" would dominate regular folk deemed inferior because they have "only" a natural human genetic makeup?

It is evident that here we are entering very murky waters. As the debate swirls around us, many are left perplexed. What is a stem cell? Is a human clone really human? Is there a difference between "therapeutic cloning" and "reproductive cloning"? Is there really anything wrong with creating "designer babies"? Is the biotech industry beneficent or dangerous?

This book will begin by addressing these and many of the other urgent questions we face as we develop capacities that only a few years ago were thought fantastical. Then we will turn to the more daunting problem of whether and how society should regulate these tremendously powerful emerging technologies.

Few areas of inquiry could be more important. The stakes are high. And we will have only one chance to get it right.

Wesley J. Smith
Castro Valley, California

CHAPTER 1

Biotech New World

. . . .

"THE THEME OF *BRAVE NEW WORLD* is not the advancement of science as such," Aldous Huxley wrote in the foreword to a new 1946 edition of his groundbreaking novel, but rather "the advancement of science as it affects human individuals." Huxley feared that science was forging "a really revolutionary revolution . . . to be achieved, not in the external world, but in the souls and flesh of human beings."[1] In other words, human biology—and *human nature itself*— could become the objects of scientific manipulation.

Brave New World portrays a future in which science is not the savior of humankind, but our conqueror. The world of the novel is one in which human society has ceased to be truly human. People are no longer "of woman born" (to quote Shakespeare); they are hatched from artificial wombs in Hatcheries run by the World State. ("We decant our babies," explains the Director of Hatcheries and Conditioning to his students early in the book.) Families no longer exist because people do not have parents. The concept of the unique individual has been virtually eradicated. The "principles of mass production" have been applied to biology: standard men and women are manufactured in uniform batches.

The resulting human sameness is more than skin-deep. Through biological predesign, human beings have long since been stripped of their free will. People are genetically engineered not only to be members of predefined social castes, but by their very design to enjoy their biologically imposed straitjackets. The hero, who is naturally born and unengineered, is viewed as a freak—the Savage—and is eventually driven to suicide after being corrupted by the stultifying antihumanity in which he is forced to live.

1

When Huxley first published his masterpiece in 1932, the technologies he described seemed unbelievable. Babies gestated in artificial environments rather than in their own mothers' wombs? It could never happen. Genetic engineering to "predestine and condition" human life toward possessing pre-selected traits and attributes?[2] What a vivid imagination! A world where applied science has alleviated all human suffering but also destroyed human aspiration and individuality? Preposterous.

Fast-forward only seventy years. Biological fatherhood is under threat of becoming superfluous: Australian researchers "may have found a way to fertilize an egg with cells from any parts of the body, rather than sperm."[3] Women also may soon become dispensable to procreation: "Doctors are developing artificial wombs in which embryos can grow outside a woman's body," *The Observer* reported. "Scientists have created prototypes made out of cells extracted from women's bodies. Embryos successfully attached themselves to the walls of these laboratory wombs and began to grow." The scientist conducting the research hopes to create "complete artificial wombs" in only a few years.[4]

Scientists are currently hard at work on learning how to "extend, enhance, or augment human capabilities far more directly, personally, and powerfully then ever before." The technologies they have developed range from the sublime to the ridiculous. Some seek the capability of merging the human mind with machine. A new social movement arising in high academia, known as "transhumanism," advocates the moral right to extend "mental and physical (including reproductive) capacities" via genetic engineering and other technologies to permit "personal growth beyond our current biological limitations."[5] The day when "a genetic vaccine that endows the user with bigger, harder muscles, without any need to break a sweat at the gym" may not be far off. Mice have already been genetically altered to develop "unnaturally muscular hind legs."[6]

The human genome has barely been mapped, but biotech companies are already developing biochips that are able to scan a person's genetic makeup in search of flaws. For both good and ill, within a few years biotechnologists will be able to map an individual's "entire human genome in 30 minutes . . . cataloging the person's genetic idiosyncrasies," leading to vastly improved diagnoses and "designer" treatments—as well, potentially, as a profound loss of medical privacy and the possibility of genetic discrimination.[7]

Biotechnology is moving at such breakneck speed that the dark vision of *Brave New World* has evolved from an imaginative literary figment and social satire into an *ideology* that sees biological transformation in an almost mystical light. A new bio-utopian mindset is emerging among futurists, bioethicists, life scientists and allied "transhumanists." These people are "committed to the process of human enhancement and self-directed evolution," which would not only embed "cultural distinctions . . . in our genetics" but eventually "increase the biological differences among human populations."[8]

The transhumanist ideology foresees a new eugenics in which parents would not only be permitted to "enhance" their offspring genetically, but in the interest of "justice" would be morally, if not legally, "obligated to do so."[9] Some proponents even foresee a distant future in which these manipulations have become so radical and widespread that biotechnologists will blur genetic distinctions between certain humans and animal species—by mixing in a hawk's DNA, for example, to improve eyesight. And others firmly believe that biological alterations will become as ubiquitous as tattooing and body-piercing are today, culminating eventually in a "posthuman" race in which some people will have such profoundly altered natures that they will be unable to procreate with others through sexual means.[10]

Posthumanity is at present a fantasy, thank goodness. But the values underlying the transhumanist movement—and the public policies likely to be implemented in attempts to reach the promised land—threaten to cause great harm in the here and now. Some bioethicists and bioscientists, for example, are pushing for society to initiate a "process of redefining ourselves as biological, rather than cultural and moral beings."[11] Undergirding this dehumanizing agenda is an almost religious belief in biotechnology as the source of human well-being. In the words of Leon Kass, chairman of the President's Council on Bioethics:

> [I]t is a way of thinking and believing and feeling, a way of standing in and toward the world. Technology in its full meaning is the disposition rationally to order and predict and control everything feasible in order to master fortune and spontaneity, violence and wildness, and leave nothing to chance, all for human benefit.[12]

If allowed free rein, where could such bio-absolutism take us? Those who have thought extensively about these issues differ. Kass warns of a "soft dehumanization of well-meaning but hubristic biotechnical 're-creationism.'"[13] He and others, such as the liberal social critic Jeremy Rifkin, predict that not only is the ideal of human equality at risk, but human life itself may be reduced to the status of a natural resource or a mere product. Yet others see these possible changes as truly beneficent, perhaps even utopian, culminating in a species devoid of imperfections. Princeton University biologist Lee M. Silver, for instance, one of the world's most enthusiastic advocates of human cloning, ecstatically predicts that bioengineering could make our distant progeny akin to gods:

> A special point has been reached in the distant future. And in this era, there exists a special group of mental beings. Although these beings can trace their ancestry back directly to homo sapiens, they are as different from humans as humans are from the primitive worms with tiny brains that first crawled along the earth's surface.... It is difficult to find the words to describe the enhanced attributes of these special people. "Intelligence" does not do justice to their cognitive abilities. "Knowledge" does not explain the depth of both their understanding of the universe and their own consciousness. "Power" is not strong enough to describe the control they have over technologies that can be used to shape the universe in which they live.[14]

Of course, it's a long way from here to there. But events move ever faster, and Brave New World has become a looming reality. In 1946, Aldous Huxley was already warning, "All things considered, it looks as though Utopia were far closer than anyone, only fifteen years ago, could have imagined. Then, I projected it six hundred years into the future. Today, it seems quite possible that the horror may be upon us in a single century."[15]

Opening the Door?

Huxley never described the events that led to the end of history as portrayed in his novel. Had he written a prequel, he might have given us a sense of what had originally induced humanity to unleash a biotechnology so powerful that it eventually resulted in a "Brave New Man ... so dehumanized that he doesn't even realize what has been lost."[16] Perhaps the ancestors of Huxley's characters were driven by an ardent desire to exercise godlike control over human health and mortality. The author might even have attempted to imagine the *one* technological discovery that began the process of "paving the road to hell with good intentions," culminating over centuries in an unintended biotechnological nightmare.

Perhaps Huxley's prophetic imagination would have led him to conceive of a scientific discovery similar to the isolation of the human embryonic stem cell (ESC) in 1998, which sparked the drive to harness and exploit nascent humans as sources for new medical treatments. This is not to say, of course, that ESC research is de facto a precursor of a morally numbed Brave New World. After all, researchers are seeking cures to serious degenerative diseases and injuries such as Parkinson's, multiple sclerosis, stroke and spinal cord trauma; they have no desire to take humanity into a dystopian future.

Still, the recent discoveries have raised very consequential moral and public policy questions that we dare not ignore. For now, I ask you to ponder a crucial question that the discovery of the human embryonic stem cell, I believe, compels us to confront head on: Does individual human life have ultimate value simply because it is human?

How we answer this fundamental question will determine the extent to which our future world becomes Brave and New in Huxley's ironical sense. If our answer is Yes, then I believe we can pursue advanced medical and biotechnological research, discovering "miracle" cures with the use of human cells, without engaging in morally degrading activities that thinkers from Aldous Huxley to Leon Kass have repeatedly warned against. If, however—whether implied through our conduct or explicitly through ideology—we answer No to the question posed, then, as Nietzsche said after postulating the death of God, everything is permitted.*

Stem Cells 101

Before tackling the hard philosophical issues and trying to connect the dots, let's pause for a moment to explore some basic facts about biotechnology.

---

*But the thought was perhaps pronounced most famously by Ivan Karamazov, a character in Dostoyevsky's *The Brothers Karamazov*.

What is a stem cell? A stem cell is the popular name for a cell that is *undifferentiated*. (Another adjective that's sometimes used to describe such cells is *immature*.) If a cell is undifferentiated, it has not yet begun to develop toward maturity—to differentiate—as one of the more than two hundred types of tissue found in the human body, e.g., blood, bone, fat, brain. Thus a differentiated cell is a "specialized cell type that carries out a specific function in the body, such as heart muscle cell, a neuron in the brain, or a red blood cell carrying oxygen to other cells in the body."[17] Until differentiation occurs, a cell is commonly referred to as a stem cell.

Where do we find human stem cells? Human embryos are the most publicized sources of stem cells. *Embryonic stem cells* (ESCs) are derived from human embryos approximately one week after their conception. At this stage of development, the embryo is called a *blastocyst*. Under the microscope, it looks like a tiny hollow ball with a cluster of cells inside it. It has an outer lining, the *trophoblast*, which develops into the placenta in the womb. Inside the lining there are between 100 and 250 cells. Some of these—no one knows exactly how many—are the stem cells that will eventually differentiate into every tissue in the body as the embryo develops.[18] ESCs are often called *pluripotent* because, theoretically, in culture they can be transformed into any body tissue.

The process of differentiation takes place in an embryo at an astonishing pace. By the fourteenth day, the *primitive streak*—the beginning of the brain and spinal cord—takes form. As gestation proceeds, some stem cells speedily differentiate into heart muscle cells, which then repeatedly divide until a working heart emerges. Generally, by the twenty-sixth day, the heart thus created is actively pumping blood—which also somehow matured, but along a different pathway, from former stem cells.

The longest-lasting stem cells are those that eventually differentiate into *germ cells* (*ova* in females, *testes* in males). These don't transform fully until about nine weeks of gestation, by which time the embryo has developed into a fetus. Consequently, embryonic germ cells are obtainable through approximately the ninth week of gestation.

But embryos are not the only source of human stem cells. Scientists have discovered that they are found in many body tissues throughout life. These stem cells are popularly known as *adult stem cells* (ASCs), even though they exist in fetuses, infants and children.

ASCs, while relatively few in number, appear to be ubiquitous. They have been discovered in bone marrow, blood, brain, fat, skeletal

muscle, esophagus, stomach, liver, pancreas, nasal tissues, hair folli-
cles—and most recently, even in the pulp of lost baby teeth.[19] Apparently these cells are part of the body's ability to heal after injury or illness
and play crucial roles in the constant regeneration of body tissues. ASCs
are often called *precursor cells* because they are undifferentiated but may
not be capable of transforming into every type of body tissue.

There are also human stem cells found outside human bodies,
specifically in umbilical-cord blood and placentas. The so-called after-
birth may provide a rich source for obtaining human stem cells that
researchers hope can be transformed into medicines. If so, we may have
a virtually unlimited supply of stem cells derivable from morally uncontroversial sources.

Why are some researchers committed to stem cell research? Scientists hope that stem cells will provide medical treatments for *degenerative* conditions in which an organ or other body system ceases to
function properly because of a breakdown or death of cells or tissues.
According to the National Academy of Sciences, tens of millions of Americans—people with conditions such as heart disease, diabetes, serious
burns, spinal cord injuries, Alzheimer's and others—have degenerative
afflictions.[20]

Parkinson's is a common degenerative disease. (Approximately
one million Americans have Parkinson's.)[21] In this brain disorder, the
patient progressively loses control over body movements as a result of
"degeneration of or damage to nerve cells within the *basal ganglia* in the
brain."[22] The condition affects the ability of the body to produce a substance called dopamine that helps the nervous system control muscle
movement. A lack of dopamine causes victims to experience progressively worsening stiffness, muscle tremors and weakness; these symptoms may become so severe that the patient becomes substantially
disabled—unable to walk and speak, perhaps losing even the ability to
eat. Death may come from complications after many years of increasing
debility and physical decline.

The illness takes a terrible physical and emotional toll. In *Saving
Milly,* the journalist Morton Kondracke wrote poignantly of the grief
and horror experienced when Parkinson's devastated his wife Milly's
health:

Milly functioned, but the life we had known before her diagnosis was over. Friends still came to call, but they weren't invited to stay over. Parkinson's disease had moved in, invading our home. It was a malevolent presence that became the preoccupation of our lives, crowding everything except our love for each other and our kids. The definite diagnosis shattered Milly psychologically. She often clung to me and sobbed piteously, sometimes several times a day, saying that her life would be terrible. "Why is God punishing me?" she cried. "I've always been good. What did I do that was bad?"[23]

While the illness can be held at bay for a period of time with medication and other therapies, until recently there was no treatment on the horizon capable of grappling with the root cause of this devastating condition. But today it appears that we may be able to harness stem cells as a cure for Parkinson's, as well as for other degenerative diseases.

This is the theory: Stem cells could be transformed from their undifferentiated state into the tissue types affected by the degenerative condition. The hope is that these cells, when injected into the body, will continue to divide and grow, eventually regenerating the damaged organs and body systems, easing symptoms and perhaps even effecting a cure. This kind of treatment, now in the experimental stages, is generically known as *regenerative medicine* because it consists of using stem cells, tissues or body chemicals to regenerate damaged structures. For example, in one experiment, mice with end-stage juvenile onset diabetes (type 1), an immune-system malady, were cured using human spleen cells.[24] The authors hope that their findings "may have implications for treatment of diabetes or other autoimmune diseases in humans."[25]

Imagine the possibilities! Brains damaged by stroke, injury or disease could be restored to proper function. Spinal cord injuries that once caused a lifetime of disability could be healed. Diabetics dependent on insulin for life might be able to wean themselves from the drug as a result of regenerated pancreases, now able to produce sufficient insulin on their own.

Why are stem cells controversial? Not all stem cell research is controversial. No one opposes regenerative medicine with ASCs or those extracted from other nonembryonic sources such as umbilical-cord blood. There is, however, great controversy over ESC research. And for good cause: the embryo is destroyed in the process of extracting its stem cells. Some opponents believe that this constitutes the taking of human life, and others, myself included, worry that destroying embryos for the purpose of harvesting their parts reduces nascent human life to the moral status of penicillin mold.

This worry is highlighted by the wording of a press release from Geron Corporation, one of the biotech companies engaging in ESC research. In announcing a purported breakthrough, Geron bragged:

> The finding greatly facilitates the development of *scalable manufacturing processes* to enable *commercialization* of hES [human embryonic stem] cell-based *products*.[26]

Meanwhile, a proposed ballot initiative in California intended to create a constitutional right to conduct ESC research and research into human cloning explicitly labels unused embryos (leftovers from fertilization procedures, a source for ESCs) as "surplus products."[27] Here embryonic human life is reduced in status to a mere commodity—a necessary legal démarche before taking the next step, namely the harvesting of its body parts into marketable items of manufacture. No wonder this technology raises moral hackles.

Many biotech boosters and members of the bioethics intelligentsia, emphasizing that embryos are not sentient and cannot feel pain, believe that the potential benefits of this research—knowledge about early embryonic development, the ability to study disease models, medical treatments—outweigh the ethical problems. Some, as will be described more fully in the next chapter, even assert (unscientifically) that a human embryo isn't really human life.

But there is little question that to condone by law the destruction of human embryos for research purposes is a critical step for us to take. Even the National Bioethics Advisory Commission (created through executive order by President Clinton in 1995), which was favorably disposed toward embryo research, asserted that "most would agree . . . embryos deserve respect as a form of human life."[28] While recommending that the federal government fund ESC and fetal tissue research, this commission proposed certain restrictions: it did *not* recommend funding the manufacture of embryos for research purposes—as opposed to the use of embryos left over from IVF fertility procedures—and it proposed laws prohibiting the buying and selling of human embryos for research.[29] And the NBAC explicitly admitted that destroying human embryos for the sake of medical progress is highly questionable:

In our judgment, the derivation of stem cells from embryos remaining following infertility treatments is justifiable only if no less morally problematic alternatives are available for advancing the research.[30]

The NBAC's challenge is actually being met by "less morally problematic alternatives" for creating a vibrant regenerative medicine sector in the form of ASCs, umbilical-cord blood stem cells, and other sources such as olfactory tissues. This removes the principal excuse not to take the NBAC's caveat seriously.

Are embryonic stem cells the same thing as fetal tissue? Although not everyone would agree, the short answer, at least from my perspective, is No. While both fetal tissue experiments and ESC research use tissues that are derived from unborn human life, and while both inject tissues for the purpose of regenerating a patient's dysfunctional body systems, let's not lose sight of a clear distinction: in ESC research, the embryo is destroyed for the purpose of harvesting its stem cells. Leaving aside the morality and propriety of abortion for the moment—admittedly something that many of my readers won't be inclined to do—fetal tissue experiments use material obtained from cadavers resulting from abortions (or natural miscarriages) that were not undertaken for the purpose of obtaining such material. Thus, it seems to me that fetal tissue experiments are analogous to organ procurement and transplantation from already dead donors. In contrast, ESC research undertakes to destroy *living* humanity.

This crucial difference established, it's worth recalling that a few years ago we heard the same promises about the potential of miraculous medical cures from the tissues of aborted babies that we hear today about ESCs. But that particular will-o'-the-wisp has proved elusive. Indeed, as we will describe later on, fetal tissue experiments in humans have generally produced negative results.

Would ESCs work better than ASCs? The jury is still out. Many biotech researchers believe that ESCs offer the better hope; or at the very least, that they should be pursued in parallel with adult cells. An article in the journal *Science* put it this way: "Most scientists ... including those who work with adult-derived cells, caution that recent [adult stem cell] advances, although promising, do not mean that adult cells can replace the need for those derived from embryos or fetal tissue. For some diseases, they say, adult cells may indeed turn out to be the better choice.

But for other applications, embryo-derived cells have some distinct advantages."[31]

According to this view, ESCs are far easier to find, given that every human blastocyst has them in an isolated packet; ASCs, in contrast, appear to be far fewer and more scattered, requiring technicians to meticulously sort them out from surrounding differentiated cells, and they are difficult to culture. In addition, ESCs are intrinsically more "active"— that is, their reproduction cycle is shorter—and theory dictates they are thus more likely to produce vigorous regeneration. (But as we'll see, this may also make them unusable if their proliferation cannot be controlled.) Better yet, proponents of ESC research claim, embryonic cells are theoretically capable of transforming into any bodily tissue (pluripotency), while adult cells may have a more limited repertoire. Finally, ESCs in their undifferentiated state are *immortal*—that is, they can be maintained indefinitely—allowing stem cell lines to be kept for use when needed.* It is worth noting, though, that some proponents of ESC research don't expect these cells to be sources of actual therapies: they think it's more likely that their role will be to assist scientists in gaining an understanding of the latent powers of adult cells.[32]

But before ESCs can be used in humans, two major problems must be overcome: tumor formation and autoimmune rejection, problems that do not appear to exist with ASC therapies. Animal studies have demonstrated the significant danger that ESCs can cause tumors. As reported in the *Proceedings of the National Academy of Sciences* in December 2000, researchers at Harvard Medical School and McLean Hospital in Belmont, Massachusetts, injected mouse embryonic cells into rats in an attempt to alleviate Parkinson's-like symptoms. Of the twenty-five rats receiving the injections, fourteen showed modest improvement, six showed no benefit, while five died of brain tumors caused by the ESCs. In other words, the treatment actually killed one-fifth of the animal

---

*The immortality argument may be weakening. James Thomson, who first extracted human embryonic stem cells, now reports that cell lines grown over several months in the laboratory "can develop genetic abnormalities, " such as gaining bits of chromosomes—changes found in "some types of cancer." This worry was confirmed in 2004 when Harvard researchers reported that seventeen embryonic stem cell lines they had created developed chromosomal abnormalities "after prolonged culture." Another 2004 study published in *Nature Biotechnology* noted the same phenomenon.

subjects, even though the researchers reduced the number of injected cells from 100,000 to 1,000—just 1 percent of the usual dose.[33] A later Parkinson's experiment showed similar results, although in this case the dosage was higher.[34]

In another animal experiment in 2003, Japanese researchers transplanted ESCs into the knee joints of mice to determine whether the cells can grow cartilage. Unfortunately, instead of growing cartilage, the cells caused tumors, "destroying the joints." The study's conclusion: "It is currently not possible to use ES cells to repair joint tissues."[35] More recently, as reported in the February 2004 issue of the journal *Diabetologia,* researchers from the University of Calgary found that insulin-producing cells obtained from ESCs caused tumors known as teratomas in mice.[36]

Tragically, a patient was apparently killed by an inadvertent injection of embryonic cells. As reported in the medical journal *Neurology,* Chinese doctors were attempting a fetal cell experiment to treat a Parkinson's patient. The patient died when anomalous tissue developed in the ventricles of his brain, perhaps as a result of stem cells differentiating indiscriminately.[37]

The second major problem confronting ESC research is the worry that patients' bodies might reject ESCs extracted from *in vitro* fertilization (IVF) embryos, just as the body tries to destroy transplanted organs. If true, this would mean that patients receiving ESC therapy could be forced to spend a lifetime taking strong drugs to suppress their immune system response.

Researchers have developed several proposed solutions to this potential problem, such as genetically engineering the cells so that they do not stimulate the body's defense system. Another approach would be to manufacture cloned embryos of patients needing stem cell therapy, and then extracting the ESCs at the blastocyst stage of development for use in regenerative therapy. This prospective approach—known as therapeutic cloning—will be defined and discussed extensively later in the book.

■ ■ ■

In contrast, ASC research has advanced to the point that regenerative therapies have been attempted in human experiments. Here are just two examples:

Dimitri Bonnville. On February 1, 2003, a nail gun accidentally discharged, driving a three-inch nail through sixteen-year-old Dimitri Bonnville's heart. The injury was severe. And then Bonnville suffered a serious heart attack, causing his heart further damage.

From the time his doctors began treatment, Bonnville's heart showed progressive degeneration. His "ejection fraction," a common measure of the heart's function, fell from a normal value of more than 65 percent to a mere 25 percent. (Ejection fraction measures the amount of blood pumped out of the left ventricle with each beat.) But Bonnville's physicians were already planning to begin a clinical trial using adult stem cells to repair damaged hearts. Being young and otherwise healthy, he seemed the perfect subject. So his doctors developed a "one patient protocol," in which they undertook to treat the teenager with his own tissues.[38]

Stem cells were first extracted from Bonnville's blood. Then the stem cells were isolated and cultured. Finally, they were injected into the coronary artery that supplies blood to the heart. A few days later, doctors noted an astonishing improvement: Bonnville's ejection fraction had risen to 35 percent, despite previous tests revealing that Bonnville had "no viable heart muscle" in the affected area. It could have been a coincidence, but the improvement seemed to indicate that the stem cells might have begun to rebuild heart function—something that had been accomplished previously with bone marrow stem cells in human experiments in France and Hong Kong.[39]

At this point it is important to emphasize that *one patient does not a new cure make.* Only time will tell whether Bonnville was rebuilding heart muscle. Moreover, even if it is proved that Bonnville's stem cells helped regenerate heart muscle, it will take much more research—with animals and in human clinical trials—before such ASC therapies will be added to medicine's arsenal as a treatment for heart disease. Still, this experiment and others appear to demonstrate that ASC therapy has the capacity to be of significant benefit in the treatment of coronary maladies. (After the Bonnville story was widely publicized, the U.S. Food and Drug Administration ordered doctors not to repeat it until extensive animal testing is performed to demonstrate its safety and efficacy.[40] However, in April 2004 the FDA did approve a human trial using bone marrow stem cells to treat heart disease.[41])

Dennis Turner. In April 2001, the California neurosurgeon Dr. Michel F. Levesque told the American Association of Neurological

Surgeons how he had treated a man named Dennis Turner for his worsening Parkinson's disease: he had used the patient's own neural stem cells. A pea-sized sample of tissue having been removed from Turner's brain, stem cells in the tissue were isolated and cultured into the millions. Then these cells were injected back into Turner's brain. One year after the procedure, the patient's symptoms were reduced by more than 80 percent—even though Turner was treated in only one brain lobe.[42]

I interviewed both Turner and Dr. Levesque about this astonishing experiment. Both told me that had Turner's disease progressed as expected, by the time of my interviews he would have required heavy medication to treat his symptoms and would likely be using a wheelchair in which he would have to be strapped. Instead, he takes only minimal medication—less than when he received the experimental treatment—and his symptoms remain mild. Indeed, Turner is thrilled that he experiences only minor trembling of the hand, and then only when he's under stress or very tired.[43] (When I called in late spring 2004 to follow up on his condition, the woman who answered his phone told me he was still doing well enough to be traveling in Africa.)

Once again, it must be stressed that one patient does not a cure make.[44] It's possible that Turner's disease would not have followed the usual progression; furthermore, there may be another reason for his apparent remission. Still, there's no denying that Turner's improved health is reason for measured optimism and further research on the use of a patient's own tissues as a potent medicine.

Those who believe that ASCs offer the best hope for a viable regenerative medicine point out that using them comes without moral baggage; and they are far less dangerous than ESCs since they do not appear to cause tumors. (There have, however, been reports that ASCs may "fuse" with other body cells, leading to cells with abnormal DNA content.[45]) And while it's true that cases such as Bonnville's and Turner's are isolated, they are becoming increasingly common; tremendous strides are being made in studies with animals and now with human patients. ASC and related therapeutic approaches are currently undergoing clinical trials or being used in the treatment of cancers, stroke, autoimmune diseases, anemia, bone and cartilage deformities, corneal scarring and skin grafts, to name a few.

Moreover, since stem cells exist in many organs and body tissues, there is probably no need for them to be pluripotent, that is, capable of transforming into every kind of tissue. (It seems, however, that a form of bone marrow stem cell may be pluripotent.) More importantly, unlike embryonic stem cells from IVF embryos that could be rejected by the patient's immune system, ASCs are genetically identical to the patient's since they are *autologous;* i.e., they are the patient's own cells. Last, it should be pointed out that ASC regenerative medicine, in the form of bone marrow transplants, has in fact been used to treat leukemia and certain cancers for many years.

So which is better? As I said earlier, the jury is still out. But the evidence is beginning to accumulate.

Are the debates about embryonic stem cells (and human cloning) really about abortion? No—and yes. Factually, abortion is quite irrelevant. But the *politics of abortion* definitely affects the debate.[46]

Pro-lifers oppose most abortions because they believe that "innocent human life" is sacrosanct from the point of conception until a natural death. Pro-choice adherents believe that laws preventing abortion (at least before fetal viability) violate a woman's human right to personal autonomy and self-determination. Thus, they support the legal right of abortion in order to ensure that pregnant women aren't forced to do with their bodies something they might not wish to do—specifically, gestate and give birth.

This dispute is irrelevant to ESC research and human cloning since applying these technologies would not force women to do *anything* with their bodies; "choice," as the term is understood in the abortion issue, is not involved at all.

What does human cloning have to do with ESC research? Many biotech researchers believe that human cloning may be a necessary adjunct to harnessing the healing potential of ESCs in order to overcome the tissue-rejection problem. The idea behind therapeutic cloning is "to generate a customized embryonic stem cell that carries the genetic blueprint of a specific patient."[47] When the cloned embryo reaches the blastocyst stage, it could be dissected for its ESCs. These, in turn, could be proliferated in culture and eventually injected into the patient. In theory, because the genetic makeup of the ESCs would be virtually identical to those of the patient, immune rejection would not take place. (There are significant moral and abundant practical problems with this technology, but we will defer that discussion until a later chapter.)

How is human cloning performed? Somatic cell nuclear transfer (SCNT), the kind of cloning procedure by which Dolly the sheep came

into being, is the primary form of human cloning that we'll be addressing in this book.[48] There are, it is said again and again, two kinds of
SCNT human cloning: one that we have already identified as *therapeutic cloning,* a term sometimes broadly used to include the cloning of
human embryos both for regenerative therapies and for use in medical
research; and *reproductive cloning*—that is, human cloning undertaken
for the purpose of bringing a cloned baby to birth.

This supposed distinction is spurious. These terms do not describe
different types of procedures. There is only one act of cloning, generally SCNT. Thus rather than describe different techniques, the terms
actually identify *different hypothetical uses* for human cloned embryos.

How does SCNT work? Generally, mammalian cloning—which of
course includes human cloning—involves the following easy-to-describe
but difficult-to-accomplish steps. Let's assume for our discussion that I
wanted to clone myself:

- A mature human egg is obtained from a woman of childbear-
  ing age.
- In a dish, its nucleus is removed.
- One of my somatic cells is selected—let's say a skin cell. (All
  body cells are somatic except for germ cells, e.g., testes and ova.)
- The skin cell's nucleus is removed.
- The nucleus from my skin cell is then inserted into the empty
  space where the nucleus of the egg used to be.
- The genetically modified egg has the full human complement
  of 46 chromosomes, namely those from my skin cell's nucleus.
- The modified egg is then stimulated with an electrical current
  or a chemical.
- If the SCNT takes, the modified egg begins embryonic development. This creates a new cloned human embryo.
- Theoretically, if implanted in a woman's womb and gestated,
  the result will be a born cloned baby.

The cloned embryo containing my DNA would be my *near* identical twin genetically. I say "near" because the *mitochondrial* DNA—
about 1 percent of the total DNA in the embryo—would consist of genetic
material that came from the egg. Thus, unlike natural identical twin
brothers or sisters, the SCNT clone made from my DNA would not be
100 percent genetically identical to me, and hence he would not actually be my "clone" in the literal sense of the term.

What are the correct terms to describe the products of SCNT? One of the main causes of confusion in the debates over biotechnology has been our inability—or refusal—to define terms accurately and use them consistently so as to keep the discussion intelligible.

This is why the President's Council on Bioethics—a panel of experts created by President George W. Bush in 2001 to advise him on biotechnology—focused at length upon the issue of definitions in its first report to the nation, *Human Cloning and Human Dignity.*[49] Despite being deeply divided over the propriety and morality of human cloning as it would be applied to medical research, the council unanimously agreed that the life brought into being via a successful SCNT cloning procedure is a "cloned human embryo."[50] This conclusion is based on sound biological analysis. The "organization and powers of this [cloned] entity, and the crucially important fact of its capacity to undergo future embryological development," the council concluded, would be "just like a sexually produced embryo."[51] In other words, the human organisms created by SCNT and those created by fertilization would be biologically the same and would develop in the same manner: "It hence deserves on *functional* grounds to be called an embryo."[52]

Since SCNT is the same procedure whether the cloned human embryo is to be used for research purposes or for reproduction, and since the embryos that SCNT produces are essentially the same regardless of how they are to be used, the council determined that the accurate terminology should be as follows:

> "Production of a cloned human embryo, formed for the purposes of initiating a pregnancy, with the (ultimate) goal of producing a child" would henceforth be known as "cloning-to-produce-children" (CPC).[53]
>
> "Production of a cloned human embryo formed for the purpose of using it in research or for extracting stem cells, with the ultimate goals of gaining scientific knowledge of normal and abnormal development and of developing cures for human disease" is to be called "cloning-for-biomedical-research" (CBR).[54]

With one exception, these will generally be the terms I use in this book. However, because CBR covers so many different uses of cloned embryos, I will also employ the term "therapeutic cloning" when describing human SCNT to be employed as a means of producing embryonic stem cells or other tissues for use in regenerative medical therapies.

Let me briefly summarize the reasons why I worry that, if we don't regulate the emerging power of biotech wisely, the discovery of the ESC could become the stimulus that sets us on the road to a Brave New World. First, as a general matter, when we countenance the destruction of embryos in order to harvest their body parts for research or medical therapies—as opposed to creating them for the purpose of helping infertile couples to procreate, during which some are lost—we cross a line of no return. We are abandoning the outlook that holds all human life to have intrinsic value simply because it is human; one subgroup of human life becomes, in effect, dehumanized and reduced to the moral status of a mere natural resource. To subject human embryos to such uses is to exploit nascent human life in the same way that we do non-human organisms.

More importantly from my perspective, the objectification of human life inherent in the ESC research enterprise would not mark the finishing line but the shot of a starter's pistol to begin the race. In just a few short years, ESCs have sparked an energetic drive by Big Biotech to legalize and legitimate research into human cloning, the essential biotechnology capable of igniting the "really revolutionary revolution" against which Aldous Huxley warned us. This accelerating effort to learn how to clone human life reliably and efficiently could well lead to the very biological transformation that Huxley predicted would occur, "not in the external world, but in the souls and flesh of human beings."[55]

CHAPTER 2

The Great Stem Cell Debate of 2001

· ·

I N 1988, JOSEPH FLETCHER, ONE OF THE MOST influential American philosophers of the last half of the twentieth century, predicted the coming of a "biological revolution" that will be no minor alteration in the way things are, no mere reform. The Harvard University professor and "patriarch of bioethics"[1] predicted that biotechnology will lead to a transformation of culture "of such a radical nature" that life scientists will one day become more powerful agents of social change than "Presidents and Parliaments and Pentagons."[2]

How will such awesome power be exercised? Using the tool of genetic screening, biotechnologists will exert "quality control" over our progeny, a process that Fletcher hoped will permit society to weed out disabled people and those he deemed inferior. "There is no such thing as a right to bring crippled children into the world," Fletcher bluntly declared barely forty years after a shocked world learned that German doctors had euthanized tens of thousands of disabled infants during World War II.[3] "If we choose family size, we should also choose family health. . . . If the State is morally justified in repelling an unwelcome invader . . . why shouldn't the family be protected from an idiot or terribly diseased sibling?"[4]

The state's right to excise an imperfect child, as if it were no more than a tumor, was only the beginning of the radical policies that Fletcher yearned to see implemented. He was eager for bioscientists to master the skills required to create "superior people" through applied genetic enhancement, and he believed that humans have an obligation to become, quite literally, self-designing; or in the Latinism he devised, *Homo autofabricus.*[5]

19

Much of Fletcher's advocacy could have been taken right from the pages of *Brave New World,* but with one crucial difference: the genetic manipulations that Huxley so urgently warned against, Fletcher wholeheartedly embraced. Not only did he extol a radical eugenics philosophy in which biotechnologists would be permitted to shuffle human genes like a deck of playing cards, but he advocated the manufacture of a slave caste of "man-animal hybrids," the moral equivalent of Huxley's fictional Epsilons: "Chimeras [part human/part animal] or parahumans might legitimately be fashioned to do dangerous or demeaning jobs. As it is now, low-grade work is shoved off on moronic and retarded individuals, the victims of uncontrolled reproduction. Should we not program such workers 'thoughtfully' instead of accidentally, by means of hybridization?"[6]

Of course, not all of Fletcher's hopes for biotechnology were so blatantly immoral; indeed, in his last book, *The Ethics of Genetic Control,* he conceded that biotechnology's staggering power could harness great good or unleash profound evil, or do both at the same time. Thus, even as Fletcher was promoting discriminatory Brave New World values, he was concomitantly urging that biotech be put to unquestionably beneficent uses that would not alter human nature or demean human dignity. For example, he hoped animals could be genetically modified so that their organs could be procured for use in transplants—*xenotransplantation*—to "relieve human beings of the risks or inconveniences of the donors' role."[7] Today, such experiments are well under way. Researchers have created herds of cloned, genetically modified pigs, which they hope will one day end the chronic shortage of transplantable organs.[8] Should this biotechnology pan out and should we find ways to ensure that animal illnesses will not be passed to human populations, the potential for xenotransplantation to alleviate human suffering is hard to overstate.

I cite Fletcher's enthusiasm for genetic engineering because he became one of our most influential thinkers by eloquently advocating "unthinkable" ideas, and because his ideas once seemed wildly futuristic. But now, less than two decades after he propounded "the ethics of genetic control," scientists have developed many of the tools necessary to bring to fruition both the light and the dark sides of Fletcher's prescient vision.

The Slippery Slope

If the cost of biotechnology's alleviation of human suffering is our acceptance of the Brave New World miasma, we will not get there in one giant leap. Rather, we will descend into the darkness in small steps, all but unaware that the shadows are lengthening. We may already have taken the first step on the proverbial thousand-mile journey. By now many of us readily accept the principle that human embryonic life, toward the end of developing new medical cures, can be sown, reaped, harvested, patented and sold—just like any other natural resource or product.

But this is only the beginning of biotech's policy agenda. In step with the widespread acceptance of embryonic stem cell research, advocates began insisting that biotech should also be permitted to engage in human cloning for biomedical research (CBR), justifying these experiments on the same grounds as they had stem cell research using IVF embryos. But cloning goes far beyond ESC research: it explicitly *creates* human life for the *purpose* of experimenting upon and destroying it. Moreover, the information gleaned in such experiments would have the perhaps unintended consequence of hastening the day of cloned human babies. Should that threshold ever be crossed, the campaign to permit genetic engineering of progeny—the fabrication of "designer babies"—would soon kick into high gear.

At this point, those who disagree with my viewpoint will probably criticize me for summoning the "slippery slope" line of argumentation, which goes something like this: If we permit controversial activity A, it will set in motion forces and changes in attitude that will lead inevitably to the even more controversial activity B. And then, before you know it, we'll be in free fall all the way down through the alphabet to activity Z.

Some believe that slippery slope arguments are exercises in alarmism. However, when it comes to the controversies dealt with in these pages, the philosopher and social critic Richard John Neuhaus has it right; asked whether he believed in the slippery slope, he replied, "Yes, like I believe in the Hudson River."[9]

The history of *in vitro* fertilization (IVF) illustrates how easily biotech can go slip-sliding away. IVF is a fertility treatment in which a woman desiring to become pregnant is *hyperovulated*—i.e., given an injection of hormones resulting in her releasing 7–10 mature egg cells during her monthly cycle, instead of the usual one egg cell. These are then extracted with a needle and mixed with the prospective father's sperm in a Petri dish. If successful, the procedure results in fertilization and the development of one or more embryos.

The embryos are nurtured in the laboratory for nearly a week, and then several of them are inserted into the woman's uterus at the appropriate time in her menstrual cycle. If implantation occurs—many embryos don't implant—the woman becomes pregnant. If all goes well thereafter, she experiences a normal pregnancy and birth. More than one million children worldwide have been born as a result of IVF since the first so-called "test tube baby," Louise Brown, entered the world in 1978.[10]

IVF was morally controversial when it was first developed. In addition to potential health concerns, some worried that society would be impotent to place reasonable limits on IVF, and that the technology would lead to an inevitable coarsening of our views about the inherent worth of the individual and society's respect for nascent human life.

Others dismissed these worries as overwrought. One notable critic of the critics was Ellen Goodman, a nationally syndicated newspaper columnist. In 1980, she wrote in support of permitting the first IVF clinic to open in the United States:

> A fear of many protesting the opening of this clinic is that doctors there will fertilize a myriad eggs and discard the "extras" and the abnormal, as if they were no more meaningful than a dish of caviar. But this fear seems largely unwarranted.[11]

Her point, one with which I agree, was that IVF could be a great boon to couples experiencing fertility difficulties, without leading inevitably to our becoming indifferent to the intrinsic worth of nascent human life. Unfortunately, that's not how things turned out.

To avoid subjecting women to the repeated rigors of hyperovulation, doctors do indeed now "fertilize a myriad eggs" and deep-freeze the extras for possible future use. Since not all embryos that are introduced into the uterus successfully implant, far more embryos are created than are actually used. As a result, hundreds of thousands of frozen embryos are maintained in cold storage, and IVF embryos are routinely dealt with in ways that were once seen as beyond the pale. Today, we do discard the "extras" and the abnormal, and we don't give much thought to the implications of our acts. Once the procedure became fairly routine, the very things that Goodman assured us would not occur quickly became commonplace and uncontroversial. And a new line has been crossed as nascent humans are now looked upon as "products" for use in ESC research.

A common defense against the slippery slope argument is for proponents to assert that they are aware of the dangers, but that our moral sense, backed by reasonable regulations, will protect against abuses. Thus, Goodman's 1980 column concluded:

> I think we should neither fund such a [IVF] clinic at this time, nor prohibit it. We should rather monitor it, debate it, and control it. We have put researchers on notice that we no longer accept every breakthrough and every advance as an unqualified good. Now we have to watch the development of this technology, willing to see it grow in the right direction and ready to say no.[12]

"No" isn't a word that many biotech researchers accepted then, and even fewer accept it now. In any case, Goodman may have been only paying lip service to the idea of monitoring and controlling this innovation; those who have labored to pry open the barn door often shrug their shoulders later on if they learn that the horse has escaped. So when I first came upon Goodman's old column, I e-mailed her, inquiring why, despite her earlier reticence and her clear statement that society should not support "every breakthrough and every advance," she has repeatedly declined over the years to actually say No as the bandwagon goes careening downhill. She courteously responded, telling me that since the date of the above-quoted article, "My lines have changed"—by which I take her to mean that she has evolved in her thinking to the point where she now readily accepts courses of action she would formerly have considered anathema.

And that's precisely how the "slippery slope" works. Despite assurances that there are limits beyond which we surely will not go, those supposedly solid walls disappear like mirages as we approach them. Assisted reproduction continues to be essentially unregulated in the United States, and now many, including Goodman, urge that we cross into a completely different moral environment in which IVF embryos aren't necessarily regarded as possible future babies, but as mere harvestable commodities. Surely, this is a dehumanizing view that would have been unthinkable to Goodman and many another supporter of IVF when the technology was first developed.

Does this mean that we should have outlawed IVF, then or now? No. But the ease with which ethical lines around IVF were redrawn should serve as a reminder of how readily we moderns accept an almost "anything goes" approach to technology. More to the point, if an innovation as benign as a medical treatment to help infertile couples has led us to the Rubicon where we have come to look upon "spare" IVF embryos as mere bundles of cells rather than as potential people, what might happen to our morality when scientists wield the godlike power to refashion our biological nature?

The Great Stem Cell Debate of 2001

In 1998, Dr. James Thomson and his colleagues at the University of Wisconsin in Madison told the world that they had successfully isolated stem cells from human embryos. Thomson's research success marked a pivotal moment in the history of science. Human embryos could no longer be viewed only as potential babies. Now, their tiniest body parts had become potentially valuable sources of medical treatments—the raw material that could lead to vast riches for the biotechnologists and venture capitalists who dominated the emerging field of regenerative medicine.

Almost as soon as Dr. Thomson published his research, the scientific community began to clamor urgently for the National Institutes of Health (NIH) to pay for ESC research. But those yearning for federal research grants faced a serious legal difficulty. It was (and as of this writing, is) illegal for the federal government to fund research that destroys embryos, a policy put in place before the discovery of ESCs and commonly known as the "Dickey Amendment" (after its original author, former representative Jay Dickey (R-AK)).[13]

President Clinton apparently wanted the NIH to fund embryonic cell research, but was blocked by the law. He turned for advice to the National Bioethics Advisory Commission (NBAC), a council of bioethicists Clinton had appointed in 1995 to advise him on medical research and other bioethical issues.[14] In 1999, the NBAC issued its report. Unsurprisingly, given the commission's makeup of prominent members of the bioethics establishment—some of whom had previously supported federally funded research on human embryos—the NBAC recommended that Congress pass an "exception" to the funding prohibition so that the government could fully finance ESC research.[15] The panel had only one substantive caveat: that the embryos used in federally funded research be limited to those remaining after infertility treatments.[16]

The NBAC's proposal would have required rewriting the Dickey Amendment, resulting in a long and heated political debate and an uncertain outcome. Instead of pursuing that politically difficult path, President Clinton looked for a less direct but still effective way around the law. He found such an approach, apparently at the suggestion of U.S. Department of Health and Human Services general counsel Harriet S. Rabb, who opined in a legal memo that the Dickey Amendment could be circumvented without congressional action: Federal funds would remain unavailable to pay for procedures involving the destruction of existing embryos and extraction of their stem cells; but researchers wishing to experiment with any resulting stem cell lines *would* be eligible for full NIH funding.[17] Thus, by seeming to follow the letter, if not the spirit of the Dickey Amendment, the Clinton administration appeared to have found a legal loophole to allow full federal funding of ESC research.

Most bioethicists, researchers, and members of the biotech industry ecstatically looked forward to the day when NIH grants would start rolling in. But then George W. Bush won the presidency and a different mindset took over. Bush was disturbed by the prospect of the government, in effect, paying to destroy human life for research purposes, and shortly after being inaugurated, he suspended—but did not rescind—Clinton's ESC funding plan.

If Bush hoped he could quietly kill federal funding by putting off the decision, he misjudged the emotional intensity of ESC proponents. Outraged at being thwarted, genuinely believing that withholding public moneys would hinder the development of wonderful new medical treatments, worrying that Bush's decision would financially damage the biotech industry, apologists mounted an energetic and effective political campaign to pressure the president to follow the Clinton approach. The result was the Great Embryonic Stem Cell Debate of 2001, a titanic ethical, scientific and political struggle that consumed the first six months of the Bush presidency and continues to shape the nation's deliberations about biotechnology to this day.

Those in support of ESC federal funding quickly formed a potent political coalition. In a brilliant stroke, celebrity disease and injury victims, whose fame guaranteed ample media coverage and fawning treatment by politicians, became the campaign's leading spokespersons. The most effective of these were (and are) movie star Christopher Reeve (quadriplegia from spinal cord injury) and television stars Michael J. Fox (disabled by Parkinson's disease) and Mary Tyler Moore (insulin-dependent diabetes); their frequent lobbying trips to Capitol Hill received high-profile media coverage followed by softball interviews on *Larry King Live* or *Oprah*.

This march of the celebrities was undergirded by politically potent and well-funded disease victim organizations, which are always accorded sympathy and support by elected representatives, the media and the public. Bioethicists, academics, biotechnology company executives and lobbyists also weighed in. Eighty Nobel Prize-winners, including James Watson, signed an ESC funding support letter drafted by Michael D. West and Robert P. Lanza, executives of Advanced Cell Technology (ACT), a company heavily invested in therapeutic cloning research.[18] The entire campaign was funded in the millions by biotech companies and coordinated by their trade association, the Biotechnology Industry Organization (BIO).

Opposing the establishment view were: the Catholic Church, most pro-life national and state organizations, the generally liberal United Methodist Church and a coalition of opponents of various political stripes who came together in a loose alliance known as the Do No Harm Coalition.[19] Physicians and scientists, including the venerable medical ethicist Dr. Edmund Pellegrino, had founded the organization. Not having the financial resources of proponents, Do No Harm operated primarily at the grassroots level, eventually turning in 140,000 signed petitions opposing federal funding of ESC research.[20]

The Great Stem Cell Debate was as much a political event as it was a scientific one. The issues raised—and the *manner* in which they were raised—showed that ESC propagandists were quite adept at mobilizing public opinion. Knowing that many Americans were uncomfortable with the notion of destroying human embryos for research, they mounted a no harm/no foul argument, repeatedly promising that only embryos already "destined to be discarded in any case" would be used in the federally funded research.[21] This argument resonated deeply with the populace—and most especially the media—by appealing to the ingrained pragmatic streak in the American character. If the blastocysts—"no larger than the period at the end of this sentence," as the slogan relentlessly puts it—were to be destroyed anyway, why not use them to help

Christopher Reeve, Mary Tyler Moore, Michael J. Fox and millions of other afflicted people?

But it wasn't necessarily true that research would be limited to left-over IVF embryos. First, under the Clinton plan, healthcare workers were to present parents with two options "at the same time": donate such embryos for laboratory study or maintain them for future attempts to bear a child, or even for another couple to do so. And in the midst of the debate, scientists at the Jones Institute for Reproductive Medicine in Norfolk, Virginia, bragged in a press release that the institute paid women between $1,500 and $2,000 apiece for their egg cells. The purpose? To use the eggs—with the providers' consent—to create embryos specifically for use in ESC research. These scientists claimed that making embryos for research purposes was just as ethical as using frozen IVF embryos, and that freshly made embryos might be superior to those thawed out of deep freeze since they would be less likely to be damaged.[22]

Various responses of ESC boosters to the institute's unabashed admission raised suspicions about their motives. "Embryos will be destroyed anyway, after all!"—was this argument more expedient than principled? Did they actually see nothing wrong with creating nascent human beings in order to kill them and dissect them? Thus Dr. Michael Soules, president of the American Society of Reproductive Medicine, with a mealy-mouthed double negative told the *New York Times* that it was "not inappropriate" to make embryos for research.[23] Meanwhile, Dr. Robert Lanza, one of the authors of the "Nobel Laureates' Letter," showed tactical shrewdness when he told the *Los Angeles Times*, "This is not good timing. They're throwing gasoline on the fire."[24] Unsurprisingly, Lanza, rather than rely on *in vitro* leftovers, intends to make "clones" for use in stem cell research.

As the Great Stem Cell Debate sharpened, it became clear that Big Biotech's ultimate agenda went far beyond gaining federal funding for ESC research from leftover IVF embryos. The simultaneous congressional debate over whether to outlaw all human SCNT cloning revealed that many enthusiasts actually believed that in order to bypass the difficult problem of tissue rejection, regenerative medical treatments would have to come from cloned embryos custom-made from the DNA of each patient.[25] Indeed, as soon as President Bush decided on partial funding of ESC research, the human cloning issue came to the fore.

Sincere or not, the assertion that unused IVF embryos are necessarily doomed is not true. As a matter of fact, a small number of IVF patients donate their unneeded embryos, not for research purposes, but to be implanted and gestated by another woman. As of this writing, forty-six babies have been born as a result of what fans call "embryo adoption."[26]

Unlike surrogacy, in which one woman agrees to gestate another woman's child, in embryo adoption the birth mother and her husband become the parents of an "adopted" embryo. While some recoil at the thought of one woman adopting another's embryo, defenders argue that bringing nascent humans to birth is more consistent with the purpose for which they were brought into being than the alternative. Less metaphysically, they point out that—in an era when "choice" rules—the entire procedure is consented to by all concerned.

The foremost organization promoting embryo adoption is Nightlight Christian Adoptions, a group founded originally to facilitate conventional adoptions throughout the United States and internationally.* Unlike almost every other pro-adoption group, Nightlight does not limit its work to babies about to be or already born; it also matches the genetic parents of embryos created during fertility treatments who want to donate their unused embryos to couples willing to become their birth parents. According to JoAnn D. Eiman, a Nightlight spokesperson, this embryo adoption project is known as the Snowflakes program because, just like each crystal of snow, "each human life is unique."[27] Snowflakes began in 1997, when human stem cells had yet to be isolated, in response to the overabundance of IVF embryos and in keeping with the pro-life philosophy that full humanhood commences upon the completion of fertilization; they presented their alternative approach as a rebuttal to the argument that since all unused embryos are essentially doomed, scientists should be allowed to use them as they see fit.

The project attracted widespread attention, including television and radio programs and a high-profile story in *Newsweek*.[28]

The Life of Jonah

I interviewed Susanne Gray of Atlanta, Georgia, about embryo adoption. "As far as I am concerned," she said, "embryo adoption is the same

---

*Based in Fullerton, California.

from a philosophical standpoint as a post-birth adoption. The only difference is that it happens sooner."[29]

Susanne has firsthand experience with the moral dilemmas following *in vitro* fertilization. Her first child was born after she used fertility drugs. Later, she and her husband, Bob, decided on *in vitro* fertilization as the means to add to their family. After giving birth to fraternal twins, they still had twenty-three frozen embryos remaining from the initial IVF treatment. (Susanne's hyperovulation resulted in an unusually high extraction of thirty-three eggs. Unbeknownst to her and her husband beforehand, they were all fertilized with Bob's sperm.) As the Grays were preparing to undergo a second regimen of implantation, they were stunned to discover that Susanne was pregnant with their fourth child.

Four children were enough for the Grays. "If we only had three or four embryos remaining, we would have tried to bring them to birth," Susanne recalls. "But when you are looking at twenty-three embryos, you could end up with an awful lot of children."

Still, despite not wanting to give birth again, the Grays didn't look upon their remaining embryos as mere blobs of tissue or collections of cells. As they pondered and prayed about what to do, they learned about Snowflakes.

Enter Cara and Gregg Vest of Hamilton, Virginia. After five years of intense effort to start a family, including unsuccessful attempts at IVF, Cara believed she would never be able to become pregnant. "I was devastated," she recalls. "I really wanted to go through pregnancy, the entire experience—from morning sickness to labor pain."[30]

She heard about Snowflakes on a radio program and contacted the organization. Snowflakes connected the Vests with the Grays. Soon, legal papers had been prepared and signed. The Vests adopted all of the twenty-three embryos and set about the process of implantation and pregnancy. Four embryos were thawed and placed in Cara's uterus. One of them implanted, but she lost the child to miscarriage.

They tried again, this time with three embryos. Two of them took. Cara miscarried one of the babies at 8/12 weeks due to chromosomal abnormalities. Finally, on May 7, 2002, Jonah David Polk Vest was born, weighing 7 pounds, 15 ounces.

The Vests and the Grays mutually decided upon an open adoption. "We wanted Jonah to know his siblings," Cara told me in a phone conversation. "We didn't want him to grow up, knowing he was adopted, knowing he had siblings, and wondering about them all of the time. So, we decided to ask the Grays into our lives."

The Grays were thrilled to know Jonah and his parents. The families have since become good friends. "We went to Jonah's christening," Susanne Gray says. "We hope to be part of all of his important life experiences."

"But isn't there a tug at your heart at seeing Jonah?" I asked.

"There is, but it is more a tug of joy and peace, not regret. Allowing the Vests to adopt Jonah made it possible for him to be born, to come into this world."

As for the Vests, "We would like Jonah to have siblings," Cara told me. "We plan to try [implantation] again."

Embryo adoption is controversial. Why, some ask, should embryos be adopted when there are so many already born children without families? Some worry that the issue could impact the running battle over abortion in the United States because of the implication that bringing embryos to birth is preferable to their destruction. Others object to the term "adoption" in this usage and believe that "embryo donation" is more appropriate, thus avoiding the message that embryos have human status. "If you can adopt embryos, how can you do stem cell research and discard them?" asked Susan Crockin, a Boston-based attorney specializing in reproductive law.[31] That's exactly the point, says Susanne Gray: she consented to her embryos being adopted to "honor the potential for human life now frozen."

It is estimated that there are about 400,000 frozen embryos in the United States alone.[32] Obviously, the vast majority of these will never be adopted, but not all are unwanted. Congress has earmarked $1 million for a publicity campaign to inform people of the potential, which perhaps will increase the number of births from this process in the coming years.[33]

Are Embryos Human?

One need not share the explicitly Christian views of Cara Vest or Susanne Gray to worry about where ESC research could lead us. Federal funding would give the moral imprimatur of the U.S. government to treating human life as a material resource, and popular acceptance of such a policy could result in our losing respect for the intrinsic value of human life.

This crucial insight was brought to my attention by Leon Kass, one of the most articulate proponents of reasonable controls on biotechnology. Kass, the chairman of the President's Council on Bioethics, told me:

> It is very rare to see scientists who think that nascent human life has any dignity worthy of being respected, whatsoever. And here, one already sees something of the dehumanizing costs that come from the familiarity of going to work on something in an experimental mode before which you no longer stand in any awe. And, if you want to see what is going to happen to the rest of us if we go down this road, you should look to see what happened to the scientists themselves. They no longer look upon early human embryonic life as something we should stand in front of in awe, because of what it can develop into. They find it unbelievable, and simply attribute it to somebody's religious superstitions, that somebody would want to protect nascent human life. [But] you don't have to believe that the embryo is a full human person to recoil from wanting to see it turned into a natural resource and there are well-meaning and sensible people—including those who are pro-choice with respect to abortion—who shudder at the thought that we would take the seeds of the next generation and turn it into a resource for the well-being of this one.[34]

The attitudes that Kass worries about have become so pronounced that some people now claim that a human embryo isn't necessarily human life; their contention is that an early embryo developing inside a woman's body is human life, but that an embryo at the same stage of development existing *in vitro*—that is, in a Petri dish—is not human life. Perhaps the most prominent spokesman for this view is Senator Orrin Hatch (R-UT), an abortion opponent who made headlines when he asserted on national television that ESC research does not destroy human life because "Life begins in the mother's womb, not in a refrigerator."[35]

If true, then geography, rather than biology, is the ultimate arbiter of when nascent life becomes human—a view that is both illogical and unscientific. As the authors of *The Developing Human: Clinically Oriented Embryology* say: "Human development is a continuous process that begins when an oocyte [egg] is fertilized by a sperm . . . ." The fertilized egg is known as a zygote, which "is the beginning of a new human being (i.e. an embryo)."[36] More to the point: "Human development begins at fertilization" with the joining of egg and sperm, which "form a single cell—a zygote. This highly specialized . . . cell marked the *beginning of each of us as a unique individual*."[37]

The authors of another embryology textbook point out that upon the completion of conception, "a new, genetically distinct human organism is formed."[38] In other words, embryos are not fungible like corn kernels; each is unique and distinguishable from every other human life that has ever lived or will ever live. If this is true of the one-celled zygote—and scientifically it's undeniable—then it's also true of the same embryo after one week, when it has developed to the blastocyst stage with more than one hundred cells, regardless of its locale.

Further evidence of how a human life begins at conception was reported in an article published in the British science journal *Nature,* which described how the human body plan "starts being laid down immediately" upon fertilization: "Your world was shaped in the first 24 hours after conception. Where your head and feet would sprout, and which side would form your back and which your belly, were defined in the minutes and hours after sperm and egg united."[39] In other words, the newly fertilized one-celled embryo is *already* a unique human life, not the "naïve sphere" or "featureless orb" that scientists once thought it to be.[40]

Whether these facts have moral significance is a different issue. As the authors of the previously quoted embryology textbook write, "The status of the early human embryo is an [ethical] evaluation rather than a scientific question, and assessment is influenced considerably by philosophical outlook."[41] But if we're going to engage in serious moral analysis, which is what society must do in these important debates, we first have to get the science right; unfortunately, the cartloads of junk biology dumped on us by the likes of Senator Hatch are intended to prevent precisely that.

This brings us to "pre-embryo," another advocacy term intended to downgrade the moral status of the early human embryo. First popularized in the mid-1980s to describe the entity that "exists for the first two weeks after fertilization,"[42] the new definition did not come about because of research breakthroughs or startling scientific discoveries, but rather as a propaganda tool.

Science tells us that biologically, there is no such thing as a pre-embryo. Thus, the authors of *Human Embryology and Teratology* place the term "pre-embryo" in the category "Undesirable Term in Human Embryology," insisting that "embryo" is the scientifically accurate, hence preferable term.[43] They explain:

The term "pre-embryo" is not used here [in their book] for the following reasons: (1) it is ill-defined; (2) it is inaccurate ... (3) it is unjustified because the accepted meaning of the word embryo includes all of the first 8 weeks; (4) it is equivocal because it may convey the erroneous idea that a new human organism is formed at only some considerable time after fertilization; and (5) it was introduced in 1986 "largely for public policy reasons."[44]

Even a few ESC and human cloning advocates acknowledge that the "pre-embryo" is a fictional concept. For example, Princeton University biology professor Lee M. Silver admits in his book *Remaking Eden:*

> I'll let you in on a secret. The term pre-embryo has been embraced wholeheartedly ... for reasons that are political, not scientific. The new term is used to provide the illusion that there is something profoundly different between what we nonmedical biologists still call a six-day-old embryo [the blastocyst] and what we and everyone else call a sixteen-day-old embryo [an embryo that has begun to develop differentiated tissues].
>
> The term pre-embryo is useful in the political arena—where decisions are made about whether to allow early embryo (now called pre-embryo) experimentation—as well as in the confines of a doctor's office, where it can be used to allay moral concerns that might be expressed by IVF patients. "Don't worry," a doctor might say, "it's only pre-embryos that we're manipulating and freezing. They won't turn into *real* human embryos until after we've put them back in your body."[45]

Such tactics and the semantic trickery that go along with them demonstrate a profound disrespect for people who are trying to grapple seriously with the ethical issues.

Another way of dehumanizing human embryos is to claim that they are nothing more than a collection of rapidly dividing cells, and accordingly, that an embryo is no different in kind from the cells of your mouth that you destroy every morning when you brush your teeth. Silver takes this position in *Remaking Eden:*

> Before joining the debate [about the morality of ESCR and human cloning], we need to establish some basic facts through a series of questions and answers:
>
> 1. Is the embryo alive? Clearly, yes.
> 2. Is the embryo human? Yes again, but so are the cells that fall off your skin every day.

3. Is the embryo human life? No. . . . The embryo does not have any neurological attributes that we ascribe to human life in the special sense.[46]

Silver calls these assertions "facts." Now, (1) is obviously true, but the
qualifying statement in (2) and the final statement (3) are at best misleading sophistries—not science at all, but ideological pronouncements.

Biologically, there is a fundamental distinction between an embryo
and a lost skin cell. According to basic embryology, "a new, genetically
distinct human organism is formed when the chromosomes of the male
and female" blend during fertilization.[47] A skin cell, on the other hand,
is not a self-contained organism. It is a part of a distinct, self-contained
organism—a chip off the old block, so to speak. Or to put it another
way, each and every one of your body's cells is a microscopic piece of
your body, but each cell is not *you*. When you were an embryo, however, that embryo was you.

An embryo has a life of its own; it is not part of another organism.
Maureen L. Condic, a professor of neurobiology at the University of
Utah, explains that an embryo functions "as an organism, with all parts
acting in an integrated manner for the continued life and health of the
body [of the embryo] as a whole." This is the "critical difference" between
cells, or a collection of cells, and a living organism. Organisms are integrated creatures. Cells are mere parts of integrated creatures.

Condic lists the functions establishing that the character of an
embryo's life is to be distinguished from that of cells:

> As distinct from a group of cells [embryos] are capable of growing, matur-
> ing, maintaining a physiological balance between various organ systems,
> adapting to changing circumstances, and repairing injury. Mere groups
> of human cells do nothing like this under any circumstances.[48]

Biologically, therefore, an embryo and a cell are utterly distinct.

Neither is Silver's third claim factual or scientific. He distinguishes
what he calls life in the "general sense" from life in the "special sense."
Life in the general sense, he claims, is "rooted within the individual cell,"
whereas life in the special sense is "rooted within the cerebral functioning
that gives rise to consciousness." He elaborates: "In human beings, life
in a special sense is localized to the region between our ears, but it lies
far beyond the level of any individual nerve cell."[49] In other words, in
order to be considered human, the being in question must possess the

ability to think; the corollary is that embryos, because they have no consciousness, possess life only in the general sense and are therefore not human beings.

The fallacies in Silver's approach are obvious. For instance, what he admits is a "highly subjective" problem is that by his definition, a human being rendered permanently unconscious by a blow on the head can no longer be considered a human life.[50]

So, is an embryo a form of human life? The biological answer is Yes. While this doesn't necessarily resolve the issues of whether the government should fund ESC research or legalize human cloning, we can at least begin to ponder them from a basis of scientific truth—not political spin, obfuscation, wishful thinking, or philosophy-masked-as-science.

"Personhood" Theory

When pushed to the wall, many promoters of ESC research will admit that an embryo is actually a human life. Then, with a quick about-face, they will claim the point to be irrelevant because human life *per se* is not what counts in determining moral worth. For them, the crucial factor—whether the life in question is pre-born or born, human or animal, organic or machine, of this world or from an alien planet—is *consciousness*. Human beings with sufficient cognitive capacities are awarded the status of "persons." Those without sufficient sentience are devalued and labeled "nonpersons." This bioethical concept is known as "personhood" theory. And for those who espouse it, personhood always trumps humanhood.

To understand the relevance of personhood theory, we must take a brief detour from our discussion of ESCs and focus for a moment on the emerging field of bioethics, a relatively new approach in philosophy that tackles issues of morality in the context of health care and biotechnology.

Bioethicists are deemed experts in matters of morality primarily because they claim to be experts. In our complex society, the views of "experts" frequently become the basis for official public policy. Bioethicists often say that their field has no overarching ideology or generally agreed-upon worldview. But this isn't true. While bioethicists certainly do argue with each other—sometimes quite vehemently—their disputes are usually about how best to apply a shared set of moral values to a given bioethical issue, rather than about what these values should be.

In this regard, bioethics discourse is more akin to Catholics arguing with Baptists than to Catholics or Baptists arguing with atheists: the disputes are generally about details and emphasis, not about divergent fundamental beliefs.

Most bioethicists believe, along with Lee Silver, that moral value and legal rights should not be based on humanhood. Indeed, many go so far as to contend that granting special status to human beings simply because they happen to be human is an act of discrimination *against animals*—an offense claimed to be as odious as racism, and labeled with the bizarre term "speciesism."[51]

To avoid the odor of speciesism, bioethicists often assert that what counts morally is whether a being is a "person," a status earned by possessing identifiable mental capabilities such as being self-aware over time or having the ability to engage in rational behavior. While the exact criteria for determining who is and who is not a person are still being debated within bioethics, the dominant view is that there are humans who are not persons.[52]

This is where the issue of personhood becomes relevant to the subjects addressed here. In bioethics generally, human nonpersons include all embryos and fetuses, since they are not capable of rational thought. But some born human beings are also denigrated as nonpersons; these include newborn infants, although some bioethicists call infants "potential persons" to avoid the consequences of depersonalization. Other designated human nonpersons include patients with advanced Alzheimer's disease, people with serious cognitive disabilities such as the comatose and those who are conscious but profoundly cognitively impaired, and people with significant developmental incapacities.*

Relying on personhood instead of humanhood as the fundamental basis for determining moral worth would certainly open the door to unlimited ESC research, with embryos created for research purposes and those left over from IVF procedures. After all, embryos obviously have no consciousness: they do not think, they do not feel; no matter what we do to them, they will never know it. Thus, supporters of ESC research

---

*It should be noted that a minority of bioethicists disagrees with personhood theory. Most of these dissenters come from an explicitly Christian or other religious perspective. Not surprisingly, these bioethicists generally have little influence within the movement as a whole.

assert, scientists not only should be able to engage in any research upon embryos deemed to advance knowledge and lead to medical cures, but also should receive federal funding to support their work.

But before we fall prey to this seductive proposition, we must consider the impact that such thinking could have on other human categories denigrated as nonpersons by bioethicists. If we can treat human embryos as mere vehicles to be exploited, patented and commercialized on the basis of their perceived nonpersonhood, could we then treat fetuses, coma victims, the profoundly cognitively impaired and other humans denied personhood status in the same manner?

Extreme though this may sound, serious discussions are well under way within bioethics to permit the enactment of just such policies. Indeed, the personhood approach is also posited as justification for other acts of human destruction and/or harvesting. For example, some bioethicists support the nonvoluntary killing of Alzheimer's disease patients, the comatose and infants born with disabilities.[53] And such policies are already in place in the Netherlands, where euthanasia is legal.[54]

The supposed lack of moral worth of human nonpersons is also advanced by some theorists as a justification for treating born people, in addition to the unborn, as so many resources to be exploited. Tom L. Beauchamp, one of the leaders of the bioethics movement, has explicitly suggested that people with poor cognitive capacities can be treated in the same way we now treat animals. He writes:

> [B]ecause many humans lack properties of personhood or are less than full persons, they are thereby rendered equal or inferior in moral standing to some nonhumans. If this conclusion is defensible, we will need to rethink our traditional view that these unlucky humans cannot be treated in the ways we treat relevantly similar nonhumans. For example, they might be aggressively used as human research subjects and sources of organs.[55]

Along similar lines, bioethicist James W. Walters thinks that "Permanent loss of self-consciousness is the crucial line between a being that possesses full moral standing and life that is without that status." Hence, people with permanent unconsciousness—an often-misdiagnosed condition—should be "able to serve as organ sources," that is, be killed and mined for their organs.[56]

This notion that we should be able to take organs from cognitively devastated patients has become so respectable in bioethics and transplant medicine that articles supporting it have now been published in some of our most respected medical journals. For example, in 1996 an article appeared in the "Department of Ethics" section of *The Lancet,* in which authorized representatives of the International Forum for Transplant Ethics argued that permanently unconscious patients should be considered already "dead" to permit doctors to lethally inject them and procure their organs.[57] Two respected physician/bioethicists went even further in a 2003 article published in *Critical Care Medicine* where they urged that the law be amended to permit doctors to kill "neurologically devastated" patients for their organs, assuming proper consent.[58]

We can now see that the moral struggle over ESC research may not be limited to the fate of embryos. After all, if one category of human nonperson can be used as a source of body parts, why not the others?

Knowing that the American people remain squeamish about destroying nascent human life, some proponents of government funding for ESC research express confidence that embryo-derived cells will certainly work as medical therapies. This encourages people to assume that such regenerative treatments are on the verge of becoming powerful tools in medicine's armamentarium, with medical miracles from ESCs just around the corner. Particularly for those with seriously ill family members, ethical and social considerations appear abstract and understandably of less consequence than the recovery of their loved ones.

This point was brought forcefully home to me in Canada when I participated in a television debate about human cloning and ESCs. After a vigorous give-and-take among the four guests on the program, the host solicited the opinions of viewers. One of the first callers challenged my perspective in emotional terms:

> *Caller:* Smith, you make some strong points, but they don't matter a damn to me. My six-year-old daughter has a terrible disability that I hope can be cured with these embryonic stem cells. I don't care what happens to the embryos. I don't care what you think about the morality of the research. I care about getting help for my daughter!
>
> *Me:* I sympathize with your family's difficulties. But let's explore this matter and see where it leads. If researchers told you it would help your daughter, would you support implanting a woman with an embryo, and then aborting the fetus after a few months to gain access to those tissues?

Caller: Yes, anything.

Me: How about gestating a fetus through the ninth month and then aborting that life to obtain tissues?

Caller: Yes, anything.

Me: How about using a newborn infant's tissues, if the infant were seriously disabled?

Caller: Yes, anything.

Me: How about using a coma victim's body?

Caller: Yes, anything.

And so it went.

Desperate people are inclined to grasp at the most morally questionable straws if they think it might help heal the object of their deepest love and devotion. While such an attitude is understandable and stems from depths of pain that no one who has not actually experienced it can fully understand, it illustrates why those who craft public policy must not allow purely emotional appeals to override their duty to engage in dispassionate analysis. This is particularly true with regard to ESC research and human cloning, which may have the potential for being developed into new medical therapies, but also for unleashing forces that could devastate the moral basis of our civilization.

My interlocutor also seemed to believe that ESC therapy was very close to becoming available to help his daughter. But at best—even if the research pans out, which is far from certain—medical treatments will not be widely available until after many years, perhaps even decades. Yet many continue to believe that regenerative medicine using embryonic tissues is at the doorstep. Why?

We need look no further than the generally poor quality of public discourse about biotechnology. While studies reported in scientific journals are generally candid about the significant difficulties that researchers face in bringing ESCs to widespread clinical application—and the length of time it will take to solve those problems—reports published or broadcast in the popular media are often far less accurate.

Christopher Reeve's advocacy in behalf of ESCs and human cloning shows how sloppy reporting of high-profile advocacy feeds public misconceptions. Before being catastrophically injured in a riding accident, Reeve was famous for his starring role as Superman. The cruel distance between this persona and the devastating consequences of his injuries—which left him completely paralyzed and dependent upon a ventilator—made Reeve an effective spokesman for increasing government funding of research into cures for paralysis.

Reeve is on fire for his cause. He travels the world giving interviews and testifying about the need for governments to back ESC research and legalize therapeutic cloning. It's safe to say that Reeve is the most famous and the most effective voice in the debate, a standing that the *National Journal* recognized in May 2004 when it named him one of the nation's ten leading "experts" in the field of "bioengineering."[59]

Reeve's zeal is admirable; less so, his frequent misstatement of the facts. For example, in a July 30, 2001 interview, the CNN White House correspondent John King asked Reeve why the focus of federal funding should not be on ASCs, since that would allow politics to be taken out of the issue of stem cell research. He answered:

> Well, that would be a big mistake because you could spend the next five years doing research on the adult stem cells and find they are not capable of doing what we know embryonic stem cells can do now. And five years of unnecessary research to try to create something that we already have would cause—well a lot of people are going to die while we wait.[60]

Most viewers, and most likely King, did not know that Reeve's statements are flat-out false. In fact, years after Reeve gave the interview, scientists still don't "know" that embryonic stem cells can become viable medical treatments. And we don't "already have" the ability to transform ESCs into medical cures. Yes, there have been initial successes in differentiating embryonic cells into various tissue types and maintaining them in culture. And in experiments with animals, differentiated ESCs have produced insulin, for instance.[61] But the popular media reporting on one experiment failed to note that this occurred in only 1 percent of the cells and that, according to expert testimony presented to a committee of the Colorado legislature, the "remaining 99 percent were a mixture of other cell types, including nerve, muscle, a few beating heart cells, and also cells which continued to proliferate."[62] This means that even with expected improvements in the science, developing ESCs as a treatment for diabetes (even if it can be made to work) remains a very long way off.[63]

In fact, rather than the steady progress Reeve imagines for embryonic stem cell research, there have been serious setbacks. As we've already seen, ESCs sometimes cause tumors, an intractable problem that does not appear to bedevil adult cells. Until the tumor issue is satisfactorily resolved, ESC therapy will remain unfit for human use. Perhaps that's why one research pioneer, John Gearhart, suggested in late 2002 that ESCs will probably not be used "in therapies." He says that "patients' own cells," e.g., adult stem cells, are "where I see the future now."[64]

Tissue rejection is the other fundamental problem that may prevent ESCs from being commonly used as a medical therapy. During the 2001 debate over federal funding, Christopher Reeve significantly downplayed the problem, which will not go away, however. For example, an article published in the May 24, 2004 *Scientific American* suggested:

> ES cells and their derivatives carry the same likelihood of immune rejection as a transplanted organ because, like all cells, they carry surface proteins, or antigens, by which the immune system recognizes invaders. Hundreds of combinations of different types of antigens are possible, meaning that hundreds of thousands of ES cell lines might be needed to establish a bank of cells with immune matches for most potential patients. Creating that many lines could require millions of discarded embryos from IVF clinics.[65]*

If true, this could doom ESCs from ever becoming a viable or widely available medical treatment.

To overcome the rejection problem, the authors of the article and much of the biotech industry suggest *therapeutic cloning,* in which each patient seeking regenerative medicine will be cloned and stem cells taken from the resulting embryos.[66] (Indeed, when Reeve acknowledges the issue, he assures us that cloning is the answer.[67] But this may be more a pipe dream than a realistic solution. As we'll discuss later, even if therapeutic cloning can be developed, there are serious, perhaps insurmountable practical and economic problems that are likely to prevent cloning from ever being widely carried out.)

In the May 1, 2001 *Time* magazine, Reeve authored a commentary touting ESCs as "the body's self-repair kit." Claiming that they are "readily available and easily harvestable," Reeve downplayed the potential for ASCs (a true *self*-repair kit), asserting, "For the true biological miracles that researchers have only begun to foresee, medical science must turn to undifferentiated [embryonic] stem cells."

---

*Nonetheless, Gearhart supports ESC research, believing it will provide valuable information about cell differentiation and other matters that will be beneficial in learning how to harness a patient's own cells for use in regenerative medicine.

Even now, years after his *Time* article, the contrary is true. Scientists still find it difficult to extract human ESCs and transform them into viable stem cell lines, and most attempts are failures. A May 2003 study found that biotechnologists still had no more than an approximate 2.5 percent success rate. Thus, out of about 11,000 embryos thought to be available for research use, the paper estimated that roughly 275 new viable ESC lines might be derived, and then "only if all of the embryos donated to research in the United States are used exclusively to create stem cells, which is highly unlikely to occur."[68] Harvard University reported in 2004 that its researchers required 344 IVF embryos—286 frozen at the 6–12 cell stage and 58 at the blastocyst stage—to derive just 17 usable ESC lines.[69] That's a productivity rate of about one stem cell line for every twenty attempts.

As to the efficacy of ASCs, as described in the last chapter and in more detail later in the book, impressive successes have already been achieved. Human trials have commenced in the United States and abroad, using adult or other nonembryonic stem cells to treat heart disease, multiple sclerosis, sickle cell anemia, immune deficiency, childhood osteoporosis, childhood arthritis, stroke, Parkinson's disease and amyotrophic lateral sclerosis (ALS), among others.[70] Moreover, ASCs are already in clinical use treating some cancers, leukemia, autoimmune diseases, anemias, bone and cartilage deformities and corneal scarring.

On March 5, 2002, Reeve testified before the Senate Subcommittee on Health that "only" embryonic cells have the potential to treat ALS, known in America as Lou Gehrig's disease. He told the senators, "I think you could bring any responsible scientist from any respectable institution to say that there is no hope at the present or projected for people with ALS, other than human embryonic stem cells."[71]

But if there was no hope either in the present or prospectively for patients with ALS other than ESCs, why would the Amyotrophic Lateral Sclerosis Association have funded $5.4 million for research projects that did not involve them?[72] At the time of Reeve's statement, there were no published reports of research successes using ESCs to treat the disease. On the other hand, scientists had shown promising results in treating ALS in animal models using umbilical-cord blood stem cells. In this study, mice with ALS-like conditions treated with such cells survived significantly longer and showed delayed onset of symptoms, compared with mice that received no stem cell therapy.[73]

In public policy advocacy, the personal often becomes the political. Thus, in his testimony before the Senate subcommittee, Reeve stated that his own case of paralysis required remyelination of his nerves, and that "only embryonic stem cells have the potential to do that."[74] Perhaps he didn't know that peer-reviewed scientific literature did not bear him out. While it is true that ESCs have shown that they can lead to remyelination in animals, the same is also true of experiments with animals using adult neural stem cells. Several studies with ASCs had, by the time of Reeve's testimony, demonstrated their ability to remyelinate spinal cords. Even better, another study had shown that implanted ASCs regenerated completely severed spinal cords in rats—an even more difficult physiological task than remyelination.[75]

This discussion raises an intriguing question: Why do many people who urgently want regenerative cures for themselves or their loved ones often seem hostile to, or dismissive of, ASC-type research successes? Or to pose the query from another angle, why do people like Reeve exhibit such an emotional attachment to therapies derived from embryos, while often seeming to discount or only faintly praise more promising approaches?

I discuss these issues more fully later on, particularly as they relate to the science establishment, the biotechnology industry and the media. But for people who are not policy makers or part of the industry, I think the answer can be found in the power of first impressions.

When stem cell research first came to public awareness in the late 1990s, most of the nonembryonic research successes had not yet been published. At the time, researchers told people that the best source of cures would be ESCs, and that religious fundamentalists, whose alleged rigidity causes them to value a tiny embryo more than a sick child, were standing in the way. The deep desire for cures, coupled with fury at those inaccurately depicted as irrational obstructionists, apparently set into emotional concrete the belief that only ESCs would suffice—all the more so since much of the information to the contrary continues to be patchily or poorly reported by the media.

Still and all, the growing weight of scientific evidence is beginning to breach the mental block caused by first impressions. Some former supporters of federal funding of ESC research now believe that sharpening our focus on ASCs and related avenues will not delay the robust development of regenerative medicine, as Christopher Reeve and others claim—and that pouring excessive resources into ESCs might do just that. Among these heterodox thinkers is James Kelly of Manitou Springs, Colorado, as committed a citizen-activist as Reeve in this intense debate, but one who is far more knowledgeable about the current state of the science.[76]

Kelly became paraplegic from a spinal injury caused by a terrible auto accident. No longer able to work as a railroad dispatcher, he has devoted 10–14 hours a day for several years to researching regenerative medicine and other treatments that could heal his spinal cord injury (SCI) and allow him to walk again.

Kelly once strongly supported federal funding of ESC research—even writing a letter to President Bush in support of such a policy. "I doubted that embryonic stem cells would cure my paraplegia," Kelly told me. "But for the sake of those with other conditions, and based on the claims I was hearing from what I assumed to be knowledgeable scientists, foundations, and institutions, I wanted to lend my support."

Then Kelly heard fellow SCI victim Christopher Reeve saying that embryonic cells and therapeutic cloning would provide the cure. Never one to ignore a possibility, Kelly decided to delve deeper into the stem cell debate and investigate whether ESCs could in fact be the key to treating his paralysis. Kelly gave his full attention to the matter, reading the many studies published in science and professional journals and corresponding or talking with many of the world's most renowned medical researchers.

Kelly emerged from this time of concentrated study sadly convinced that the likelihood that ESCs could cure spinal cord injury was "more hype than hope." Worse, he worried that the intense political focus on ESC research and therapeutic cloning could result in slowing down the pace of medical advances for SCI and other degenerative conditions. "The public advocacy surrounding these fields is so overblown that it threatens to divert substantial research resources away from potentially efficacious sources of cures for SCI," he told me. "We don't have to travel down long paths that will probably not lead to any cures simply for the sake of leaving no stone unturned. What we have to do is use our limited resources efficiently. Money spent on ESC research and cloning is money that cannot be spent on [developing] adult stem cells. And that means that the cures that I believe can be found will be slower in reaching the patients who need them."[77] As for his own hopes of walking again, Kelly is most excited by research being conducted by Dr. Carlos Lima in Lisbon, Portugal, who has reported optimistic results in early human trials by harvesting a patient's adult olfactory tissues and injecting them into the spinal cord.[78]

The President Decides

In 2001, as spring turned to summer, the Great Stem Cell Debate came to a head. The political forces backing federal funding for ESC research seemed to be running with a favorable wind. In the United States Senate, 60 senators signed a letter proclaiming their intention to vote in support of full federal funding. In the House, 260 representatives were similarly disposed. President Bush consulted closely with advocates on both sides of the issues: scientists, scholars, physicians, ethicists, religious leaders and members of his cabinet.

In August, the president was finally ready to announce his decision. In his first nationally televised policy address, George W. Bush told the nation, "Scientists believe that further research using stem cells offers great promise that could improve the lives of those who suffer from many terrible diseases." Further, he acknowledged that biotech researchers had stated their opinions that embryonic stem cells offered the "most promise" for developing future treatments because of their natural ability to become all types of human tissue and that "rapid progress in this research will come only with federal funds."

On the other hand, the president pointed out, ESCs were not the only option. "Many patients suffering from a range of diseases are already being helped with treatments developed from adult stem cells." Moreover, extracting stem cells from embryos kills them, and for many, this raises profound moral issues.

The president summarized the moral dilemma in which he—and we—find ourselves:

> At its core, this issue forces us to confront fundamental questions about the beginnings of life and the ends of science. It lies at a difficult moral intersection, juxtaposing the need to protect life in all its phases with the prospect of saving and improving life in all its stages. As the discoveries of modern science create tremendous hope, they also lay vast ethical mine fields. As the genius of science extends the horizons of what we can do, we increasingly confront complex questions about what we should do. We have arrived at that brave new world that seemed so distant in 1932, when Aldous Huxley wrote about human beings created in test tubes in what he called a "hatchery."

Yes, the president agreed, science and medical progress are important. However:

> While we must devote enormous energy to conquering disease, it is equally important that we pay attention to the moral concerns raised by the new frontier of human embryo stem cell research. Even the noblest ends do not justify any means.

The president then told the nation that after contemplation and prayer, he had decided on a compromise. Embryonic stem cell lines that already existed, made from embryos that had already been destroyed, would be eligible for federal funding. He also decided to increase federal funding of adult and other nonembryonic stem cell research. And he established an advisory panel—the President's Council on Bioethics—to "monitor stem cell research, to recommend guidelines and regulations, and to consider all of the medical and ethical ramifications of biomedical innovation."[79]

Like most compromises, this one pleased some and angered others. The American Council of Catholic Bishops opposed the president. Other opponents of ESC research supported him, seeing his middle way as more defensible politically than a total ban on federal funding. The research community expressed similar mixed feelings. Some were pleased that funding would be forthcoming to pursue the research. Others groused that restricting funding to existing lines would dampen progress, a complaint that has grown louder as time has passed.

By September, some in Congress were threatening to mount a political effort to compel full federal funding. Then came the events of September 11, 2001. The nation's attention, and that of its political leadership, became fixed on more urgent matters.

President Bush's compromise finessed the ultimate question about the proper limits to place around biotechnological research involving nascent human life, and the Great Stem Cell Debate of 2001 was only a preliminary bout, not the main event that is still to come. But the stakes in the struggle over humanity's future, lurking just beneath the surface throughout the controversy, soon became vividly clear. It was only a moment before biotech advocates—the same ones who had promised that all they needed to provide medical miracles was access to leftover IVF embryos—increased their demands. Access to existing embryos was not enough: they also needed a legal license to create new embryos via SCNT human cloning.

We are still embroiled in this crucial debate—perhaps the most important ethical controversy of our time. How we resolve the conundrums posed by human cloning could well determine whether our traditional conception of individual human worth—indeed, our very definition of humanity itself—will remain honored, or whether we are on the fast track to a Brave New World.

CHAPTER 3

Reproduction As Replication

· ·

O

N DECEMBER 27, 2002, BRIGITTE BOISSELIER, a rather odd-looking woman with badly bleached hair and a French accent, held a press conference in Hollywood, Florida. Grinning broadly, Boisselier proclaimed that the first human cloned baby, allegedly a girl named Eve, had been born.[1] Her announcement sparked frenzied headlines around the world.

It was almost certainly a hoax, although to this day Boisselier insists that her activities continue to result in cloned babies.[2] Yet even though she presented no evidence of any kind supporting the claim, though she had not submitted her work for review by respected scientists, the world's media erupted into an all-out feeding frenzy as if a Nobel Prize-winning scientist had made the announcement.[3]

Boisselier had a distinct aura of the fanatic about her. She claimed to be the head of a company called Clonaid, which touted itself as "the first human cloning company."[4] In reality, the *Boston Globe* reported, Clonaid was an empty shell—with no address, no board of directors and only two employees.[5] It was founded by and remains a front for the Raelians, an atheistic science cult that believes visitors from outer space manufactured, via cloning, everything that lives on earth.[6] Founded by a Frenchman named Claude Vorilhon, who now goes by the name Rael, Raelians believe that extraplanetary beings instructed Vorilhon to establish an embassy to welcome our life-founders back to Earth. According to their ideology, human immortality will be achieved in this present physical life through cloning and the downloading of minds into the brains of successor clones.[7]

In the weeks after the announcement, Rael and Boisselier became international celebrities thanks to a kowtowing media jockeying for interviews; Connie Chung went so far as to accede to Rael's demand that he be called "Your Holiness" on her CNN show. It wasn't until after Boisselier announced that "Eve" and her mother would not be made available for verifying tests that the media finally decided to call a hoax a hoax.

At the time this pseudostory broke, no cloned monkey had yet been birthed and no cloned human had even gone past the first few cell divisions, making it highly unlikely that Boisselier could have succeeded where those with rather more scientific credibility had repeatedly failed. Up till then, the most successful human cloning experiment had been conducted by Advanced Cell Technology (ACT), which made headlines when it announced that it had created a cloned human embryo through somatic cell nuclear transfer (SCNT) and had kept it growing until it died at the six-cell stage.[8] Even that claim, published online, was angrily criticized by some scientists—one even called it "a complete failure"—because human egg cells can be induced to divide briefly without fertilization or SCNT.[9] (Subsequently, South Korean experimenters produced human cloned embryos and were able to maintain them up to the blastocyst stage of 100–200 cells.)

It's easy to poke fun at gullible news peddlers for sensationalizing the Raelian hoax. But if the story was grossly overblown, perhaps it was because a multitude of people feel uneasy about bringing human clones to birth; they are aware that crossing such a Rubicon would be an epochal moment in human history. A world in which a person could be made "asexually"—that is, without the contribution of two parents' genes—would be radically different from our current existence, where everyone is the result of normal fertilization.

Like Frankenstein's monster being animated by a bolt of electricity, cloning had received a tremendous boost almost six years earlier, when in February 1997 the Roslin Institute of Edinburgh, Scotland, announced the birth of the world's first mammalian clone brought to birth using the SCNT process: Dolly, the lamb.*

---

*A cultural note: Dolly was named after the big-busted country-and-western star Dolly Parton. The reason? The scientists had used a mammary gland cell in the SCNT process.

Dolly's birth changed everything. In *The Second Creation,* Ian Wilmut and Keith Campbell, her creators, recall the tremendous emotional response that the animal's birth evoked throughout the world:

> Dolly's impact was extraordinary. We expected a heavy response ... but nothing could have prepared us for the thousands of phone calls (literally), the scores of interviews, the offers of tours and contracts, and in some cases the opprobrium, though much less of that than we might have feared. Everyone, worldwide, knew that Dolly was important. Even if they did not grasp her full significance, (and the full significance, while not obvious, is far more profound than generally appreciated), people felt that life would never be quite the same. And in this they are quite right.[10]

From the start, there was much confusion about what cloning was and what it could do. Dolly's appearance on the scene was the beginning of a huge argument, which is only warming up, over the proper controls to place on biotechnological enterprise—and indeed what it means to be human in a biotech age. Wilmut and Campbell did not exaggerate when they wrote, "In the twenty-first century and beyond, human ambition will be bound only by the laws of physics, the rules of logic, and our descendants' own sense of right and wrong. Truly, Dolly has taken us into the age of biological control."[11]

In the immediate years after Dolly's entrance, many animal species were cloned and brought to birth—including pigs, cows and mice. But after the initial brouhaha, few people apart from some environmentalists seemed to think seriously about the implications. It was not until the realization dawned that the first human cloned baby could be just a few years away that the debate about ethics went into high gear. On the pro-cloning side is Big Biotech: the biotechnology industry and its lobbying arms; the medical science and research establishment; the bioethics intelligentsia and futuristic dreamers, who have footholds in both academia and the marketplace; and some patient organizations. However unlikely this disparate coalition may seem, they agree on strategy: their current position is that bringing a cloned baby into the world should be illegal—for now—but that researchers should be allowed to create human clones as a means of advancing regenerative medicine and studying embryonic development.

Opinion polls show that the majority of Americans, and much of the world, oppose cloning human life.[12] Activists working by political means to criminalize human SCNT include environmentalists, some feminists, most pro-life organizations and many churches; their arguments that human cloning is morally wrong, and that learning how to clone human embryos would inevitably result in cloning for reproductive purposes, have helped persuade Australia, Norway, Germany, France, Canada, Taiwan and other countries to enact legal bans against human SCNT. (As of this writing, the United States has no federal policy governing human cloning.) Four countries explicitly permit human SCNT for medical research purposes: the United Kingdom, Israel, Saudi Arabia, Singapore and the People's Republic of China.

Cloning-to-Produce-Children Is Not Safe

Given that cloning is such a contentious moral and political controversy, it may seem surprising that almost all human cloning advocates agree with opponents that cloning-to-produce-children (CPC) should be prohibited: it is simply too unsafe. But this is so obvious, so uncontroversial scientifically, that all but a few moral outlaws and rogue scientists—such as Severino Antinori, an infertility specialist from Italy, and Panayiotis Zavos of the University of Kentucky, and of course the Raelians—agree that it would be monstrous to attempt to bring a cloned human baby into the world at this time.

The history of Dolly illustrates the problem. It took 277 cloning attempts to bring her to birth, before which 276 cloned sheep embryos had been made via SCNT and implanted in ewes; altogether, only thirteen resulted in pregnancy, and only one of these resulted in a live birth.[13]

At first Dolly appeared to be a healthy lamb. But she soon developed disturbing health problems. In 1999, at the youthful age of three, she began to show distinct signs of premature aging. An investigation revealed that her genetic age was probably older than her actual years—she may have been born the genetic age of the ewe whose DNA was used in the SCNT cloning procedure.

This may be why: Dolly was born with shorter-than-normal cell *telomeres*.[14] These are the protective tips on the chromosomes of all mammals. Over time, indeed with each cell division, our telomeres grow minutely shorter. (The human body continually replaces its cells and tissues.) This seems to be part of the aging process. As telomeres grow ever shorter, health problems associated with aging begin to occur, such as cancer, arthritis and other degenerative diseases.

Dolly developed obesity at a very young age. She also suffered from early-onset arthritis. Finally, she developed progressive lung disease and had to be euthanized. Sheep generally live for eleven or twelve years; Dolly died at age six.[15] Stuffed, she is now on display at the National Museum of Scotland.

Dolly's health difficulties are only the tip of the iceberg of the problems that have developed with animal cloning, all of which are relevant to the propriety of attempting to bring a human clone into the world. Even when SCNT successfully results in a cloned animal embryo, few have been brought to birth successfully. As of April 15, 2002, a review of the scientific literature found a very low live birth success rate in animals. In one study, for example, only 5 live births resulted from 613 cloned mouse embryos. In a pig cloning program, there were 5 live births out of 72 implanted embryos, a 7 percent success rate. The famous birth of the first cloned cat came about after 87 attempts.[16]

In an experiment in 2003, four cloned piglets were born. Of these, one died within days. Within six months, all the remaining pigs collapsed and died of heart attacks. One of the stunned researchers who had created them coined a name for the premature deaths of cloned animals: "adult clone sudden death syndrome."[17]

But, some people will say, science has only been cloning animals for a few years. Surely, these early dire statistics will improve. Perhaps— but perhaps not. Despite several years of mammalian cloning, the success rate has not improved.[18] It seems that the safety problems associated with cloning involve far more than unperfected technique; the cloning process *itself* may lead to biological defects that scientists may not be able to prevent consistently.

The problems with mammalian cloning arise at the minute level of the genes. Rudolf Jaenisch, the noted cloning expert at MIT and the Whitehead Institute for Biomedical Research in Boston, analyzed the genes of cloned mice and found that some of them expressed abnormally in the placenta, meaning that "the signals telling the fetus how to develop are scrambled," causing miscarriage in most cases. But even mice that are brought to birth and appear normal often have serious defects that may cause "premature death, pneumonia, liver failure, and obesity in aging cloned mice."[19] In monkeys, the mutations are so damaging that cloned monkey embryos have been described as a "gallery of horrors." As of this writing, no live cloned monkey has ever been born.[20]

The medical journal *The Lancet* reported in March 2003: "Many researchers are concerned whether any [mammalian] clone is normal."[21] No wonder Jaenisch told the Science, Technology and Space Committee of the U.S. Senate, "Our experiences with animal cloning allows us to predict with a high degree of confidence that few cloned humans will survive to birth and of those the majority will be abnormal."[22] (Some readers may be reminded of one of the *Alien* movie sequels in which the Sigourney Weaver character learns she is a clone when she stumbles upon horribly deformed earlier versions of herself.)

It appears that cloning a human would "probably damage more than 1000 of [the embryo's] genes," and that "even normal-looking clones would harbor hidden health problems."[23] As a consequence, "Cloning is far too dangerous to try on people," wrote science writer Rick Weiss in the *Washington Post*. "Cloning, it turns out, is a serious health risk usually resulting in death for the clones.... Moreover, as cloners expand their efforts to a growing variety of animals, including cows, goats, sheep, and mice, it's becoming clear that the problem is not simply one of beginner's bad luck." Apparently it is caused by "a disruption of a genetic mechanism known as 'imprinting,' which is nature's way of ensuring that a baby has two parents."[24]

In normal reproduction, we receive two different copies of every gene—one from our father and one from our mother. Each copy of the gene has a specific biological task or tasks, instructing the cells of the body to perform their biological functions. The term "imprinting" refers to whether a particular biological function is controlled by the gene inherited from the father or the mother.[25]

Researchers have discovered that proper imprinting is crucial to healthy embryonic and fetal development. Here, an analogy might be a railroad crossing with automatically controlled warning barriers. The system depends on a program that tells the barriers to stay up or down at the appropriate times. If a defective program sends the wrong commands—expresses itself incorrectly—the unfortunate results will be traffic jams or, worse, catastrophic accidents. Likewise with the imprinting function: if it goes awry and the affected genes turn on and off in incorrect sequence, the result will be abnormal placental or embryonic development, miscarriage or lasting health problems.

Since somatic cell nuclear transfer (SCNT) involves the creation of an embryo without the usual merging of chromosomes from father and mother and the myriad other interactions that occur when sperm meets egg, perhaps we shouldn't be surprised that the imprinting function in the early cloned embryo is almost always defective. We must also remember that the SCNT procedure involves the forced merging of parts from two different cells—the egg and somatic cell nucleus from the clone donor—that were not designed by nature to be fused in this way. Moreover, the two cells used in the process will almost certainly be damaged in the process of extracting their respective nuclei. As Stuart Newman, professor of cell biology and anatomy at New York Medical College, told me:

> You could characterize cloning as a simple thing but that obscures a lot of damage that you are doing during the procedure. You are taking a cell that has been damaged by ripping its nucleus out of it [the egg] and taking the nucleus of another [somatic] cell [from the clone donor] that has been damaged by breaking it apart—these are two fragments of damaged cells—and hoping that they can produce a presentable member of the human species. But why should it? You are attempting to do in the laboratory what evolution has not produced: fragments of cells used as a way of making new organisms. If these fragments of cells are ever able to make new organisms, they are inevitably going be defective organisms.*

So the defects found in cloned mammals may simply be a matter of breaking Humpty Dumpty and not being able to put him fully back together again.

Worse, not only would the clones very probably be defective, but because SCNT often leads to enlarged fetuses or placentas in animals, implanting a woman with a cloned human embryo could entail serious risks for the mother, perhaps even endangering her life.[26] (In one cattle cloning study, *one-third* of the implanted cows died during their pregnancies.)[27] Ian Wilmut spoke for virtually the entire planet in 2001 when he unequivocally denounced as "extremely cruel" the reported plans of renegade fertility specialists Severino Antinori and Panayiotis Zavos to implant cloned human embryos and attempt to bring them to term.[28] It is hardly surprising that almost all scientific organizations (such as the

---

*Stuart Newman, interview with author, September 25, 2002. Dr. Newman is a board member of the Council for Responsible Genetics (CRG), which has an informative website at www.genewatch.org.

National Academy of Sciences), the biotechnology industry, professional medical and bioethics journals, as well as most individual scientists working in the field agree that cloning-to-produce-children (CPC) should be illegal—at least until the safety issues are resolved.[29]

The best discourse on the safety problems associated with CPC may be *Human Cloning and Human Dignity,* published in 2002 by the President's Council on Bioethics. The council divided bitterly (10–7) over whether there should be a moratorium against cloning-for-biomedical-research (CBR). But it voted unanimously to recommend that the United States permanently illegalize attempts to bring forth a cloned human baby.

But the council took the matter a step further than the usual analyses published in bioethics and scientific journals about the lack of current safety. Not only is human cloning not safe now, the council decided, but also *"There seems to be no ethical way to try to discover whether cloning-to-produce-children can become safe, now or in the future."*[30] According to the unanimous council, citing animal studies such as those mentioned above, we can never know the actual level of risk to a woman in attempt-' ing to gestate a cloned embryo to term. Considering that this is *human research* governed by strict ethical and legal standards, such experiments should never be conducted:

> If experiments to learn how to clone a child are ever to be ethical, the degree of risk to that child-to-be would have to be extremely low, arguably no greater than for children-to-be who are conceived from union of egg and sperm. It is extremely unlikely that this moral burden can be met, not for decades if at all.[31]

To better understand the council's concerns, consider the experimental track that would have to be undertaken to learn how to bring a cloned human baby "safely" into the world. Dr. David A. Prentice, a biologist and adult stem cell researcher, is one of the world's foremost experts on the science and ethics of human cloning. An opponent of human cloning, Prentice has traveled the world testifying before governmental bodies, lecturing in public forums and participating in scientific and bioethics symposia about these matters. He described for me the intense trial-and-error methodology that would be required in order to learn how to engage in CPC safely:

Scientists would have to clone thousands of embryos and grow them to the blastocyst stage to ensure that part of the process leading up to transfer into a uterus could be "safe," monitoring and analyzing each embryo, destroying each one in the process. Next, cloned embryos would have to be transferred into the uteruses of women volunteers. The initial purpose would be analysis of development, not bringing the pregnancy to a live birth. Each of these clonal pregnancies would be terminated at various points of development, each fetus destroyed for scientific analysis. The surrogate mothers would also have to be closely monitored and tested, not only during the pregnancies but also for a substantial length of time after the abortions.

Finally, if these experiments demonstrated that it was probably safe to proceed, a few clonal pregnancies would be allowed to go to full term. Yet even then, the born cloned babies would have to be constantly monitored to determine whether any health problems develop. Each would have to be followed (and undergo a battery of tests both physical and psychological) for their entire lives, since there is no way to predict if problems [associated with gene expression] might arise later in childhood, adolescence, adulthood, or even into the senior years.[32]

The absurdity, not to mention the immorality, of treating human life as mere raw material to be experimented upon and tossed aside if it doesn't happen to have a successful gene configuration should be self-evident. It would permit us to turn these unfortunate people into lifelong medical experiments. Little wonder, then, that the report of the President's Council was so unequivocal in its condemnation.

But some bioethicists and biotechnologists are not deterred: Seduced by a vision of scientific pursuit without limits, ideologically predisposed to support futuristic scenarios of virtually unlimited "choice" in the ways in which individuals procreate (or replicate), and anxious to get on with the task of species "enhancement" through genetic engineering at the earliest practicable date—which for a variety of reasons would require extensive human cloning—these extremists urge that we steam full-speed ahead to create cloned human children. Thus Robert Edwards, for instance, whose work with Patrick Steptoe led to the birth of Louise Brown, the first baby born as a result of IVF treatment, suggested that one day society will regard human cloning as being merely a first cousin of IVF. Eschewing all moral objections to the practice and apparently indifferent to the horrendous process that would be required to master it, he told the *Sunday Telegraph* of London: "I would not object to [the birth of a cloned child] provided all embryos after cloning are as normal as those after normal conception."[33]

Ronald M. Green, director of the Dartmouth College Ethics Institute and an active supporter of ESC research who has served on government committees and is currently a member of the ethics committee of the human cloning biotech firm Advanced Cell Technology, also supports eventual permission to engage in CPC. Green suggests that "caution and not necessarily prohibition" should be the regulatory byword regarding CPC. While he admits that the first generation of cloned children would be likely to face significant life difficulties, he also foresees "novel satisfactions" to being one of the first human clones. "As cross-generational twins," Green wrote in *Scientific American,* "a cloned child and his or her parents may experience some of the unique intimacy now shared by sibling twins. Indeed, it would not be surprising if, in the more distant future, some cloned individuals chose to perpetuate a 'family tradition' by having a cloned child of themselves when they decide to reproduce."[34]

John A. Robertson, a bioethicist and law professor at the University of Texas School of Law, is already promoting a right to clone as part of what he sees as an almost absolute right to procreate. If a married couple is infertile, he writes, he sees no arguments "sufficient to justify infringement of [their] presumptive right ... to rear genetically-related children" through reproductive cloning.[35] He also foresees lesbians gaining the right to procreate through cloning in order "to avoid male involvement," which would set personal predilection on an equivalent moral plane as the problem of physiological infertility.[36] Moreover, he anticipates that as society slouches toward greater acceptance of cloning, it will find new uses and purposes for this technology.

Yet another human cloning enthusiast is University of Alabama bioethics professor Gregory E. Pence, author of *Who's Afraid of Human Cloning?* Pence supports the technology and its use in genetic engineering and bitterly criticizes some of its most prominent opponents, such as Jeremy Rifkin and Leon Kass, as "false prophets of doom."[37] Specifically, Pence claims that because Rifkin and Kass allegedly opposed IVF when that technique was first emerging, and since IVF has not undermined society or led to children born with disabling birth anomalies, their current qualms should be disregarded as so much paranoia.

I interviewed both Rifkin and Kass about these and similar criticisms frequently expressed by biotechnology boosters. Interestingly, these two public intellectuals have little in common other than their opposition to human cloning. Kass is a traditionalist whose politics and ethics definitely lean toward the conservative side. He is a strong defender of marriage, for example, and the importance of children to family and society. Rifkin, on the other hand, avowedly walks on the far-left side of the road—even supporting the granting of rights to animals, which he sees as "our fellow creatures."[38] He was an early biotechnology Paul Revere who raised the alarm in 1976 when he organized the first public protest against out-of-control biotechnology at a National Academy of Sciences meeting. A year later, his *Who Should Play God? The Artificial Creation of Life and What It Means for the Human Race* (co-authored with Ted Howard) warned against the dangers of an uncontrolled "biotechnological revolution."[39]

Scientists and research establishment types castigated the book in its day as so much alarmism. Yet the book was prescient. Most of Rifkin's warnings—about the development of hybrid species, the emergence of a new eugenics using genetic engineering to "improve" the human race, and the impact of commercializing biotechnology—have either come true already or are on the Brave New World drafting board.

Rifkin distinguishes between *soft* and *hard path* technologies. He opposes the latter, which include the genetic engineering of food, mixing the genes of different animal species (transgenic animals), and human cloning. "The hard path is based solely on utility," he explained. "The soft path is based on intrinsic value. They lead to completely different applications."

The soft path does not admit human cloning, which "redesigns" the natural order and is profoundly contemptuous of our intrinsic nature. Rifkin does, however, support adult stem cell therapies, because "You are not trying to make the genes, the proteins, the cell lines do something they wouldn't normally do. You are just nurturing them to complete the task that they have already been assigned in the long evolutionary scheme of things."

As to IVF, Rifkin sees it as acceptable because it mimics nature, but does not alter it or change the biology of the human individuals who come into being through the procedure. However, Rifkin *does* worry that "IVF may be an essential step in the progression toward removing the [reproductive] process from the body," leading us away from respect for the intrinsic value of human life and along a stone-hard path that would perceive progeny as malleable commodities that could be manufactured, patented and remade. "If commercial forces gain control of human reproduction," Rifkin warns, "we will apply engineering and assembly line values to the creation of children: quality control, predictive outcomes, quantifiable standards of measure, efficiency and utility."[40]

Kass reacted strongly against the charge that his concerns about *in vitro* fertilization were rooted in an antiscience outlook. "Yes, I raised questions about IVF before the first test-tube baby, Louise Brown, was born," Kass told me. "But no one goes back to read those original arguments of mine; they simply recycle the stripped-down conclusion, 'he was opposed to IVF.'"

"What worried you?" I asked him.

"Part of my early objection had to do with experimenting on the unborn; was IVF safe enough to use in humans? The fact that it was tried and turned out not to be terribly unsafe does not prove that it was in fact ethical to have gambled with the future child in the first place."

While Kass's worst fears about the safety of IVF have, thankfully, not been realized, it's worth noting that children born as a result of IVF treatments do seem to have more health problems than children conceived naturally. Researchers have found that children conceived via IVF are more likely to have certain birth defects and may be predisposed to Wilms' tumor and other cancers.[41]

More to the point, Kass's "major objections" to where IVF could lead are in the process of being realized. "I worried that once one had human life in human hands, there would be other uses for this technology beyond intramarital treatment of infertility; surrogacy, embryo research, cloning. My forecasts have not been proved wrong. Neither has my prediction of the way in which the justification of the more innocent uses would come to serve also as justification for the less innocent uses."[42]

Indeed, as we have seen, leftover IVF embryos *are already* being reduced in moral status so that they can be exploited as if they were nothing more than a harvestable crop. Biotechnologists and their allies *have already* invoked the medical potential of ESC research to argue that they should also be allowed to engage in human cloning-for-biomedical-research. The step-by-step advance toward Brave New World that Kass has repeatedly warned against is well under way.

Legalization of CBR would lead directly to cloning-to-produce-children, which would in turn open the door to genetic engineering and attempts to redesign the human genome. Already, lawyers and bioethicists are arguing that all of these activities are protected by the fundamental liberty to procreate guaranteed by the U.S. Constitution. Clearly, Rifkin and Kass are not paranoid alarmists about where the road leads, as their opponents often contend. Rather, they are prophetic voices whose warnings we ignore at our peril.

Moral Reasons to Oppose Cloning-to-Produce-Children

When comparing the arguments for and against permitting CPC, I have noticed a striking difference between its supporters and those who believe that engaging in "asexual reproduction"—as cloning is sometimes called—is morally wrong under all circumstances. Supporters of human cloning focus primarily on the rights/desires/needs of the would-be cloners. Opponents, on the other hand, generally focus on the negative impact cloning could have on the child and ultimately upon society as a whole.

Science reporter Rick Weiss finds good examples of the former mindset in a May 2002 *Washington Post* story. Profiling people actively seeking to create cloned babies for themselves, the article is appropriately titled "Free to Be Me." One of the women in the story is infertile. She wants a child, but "not a child made from a donor egg provided by someone she doesn't know. Not one adopted from halfway around the globe. She wants a baby genetically related to her. And if that means one who's genetically identical, then so be it."

Then there is a grieving mother of a dead teenager named Emily. In her bereavement, she has become "obsessed with the idea of cloning a girl from some of Emily's cells." She tells Weiss, "I don't understand people who want to clone themselves [but] I'd trade everything I have today just to have Emily back." Knowing intellectually that the cloned child would not actually *be* Emily, she says, "If I could have a child with her predisposition to life, her humor, but have her grow up in this new life I've created for myself, which is much better now [than when Emily was alive]. I'm married to an attorney. I have all kinds of things now."

Another would-be cloner is a man with cystic fibrosis who wants to make a copy of himself with the CF gene removed. "In some respects," he tells the *Post* reporter, "it would give me a second chance at life without CF. It wouldn't be me, but it would be very similar to me."

We should have sympathy for these sad people. But it's striking how uniformly their desire to clone is based exclusively on a "me" mindset. Even Weiss, who often boosts biotechnology in his stories and seems tolerant of people's right to clone—for him it's a matter of upholding their American right to engage in the "pursuit of happiness"—notices the narcissism his subjects display. "Perhaps what disturbs some ethicists more than anything," he comments, "is the literal self-promotion of it all. Cloning, after all, is about the ascendancy of the individual, the chance to propel one's genetic self into the future, undiluted by another."[43]

This same "it's all about me" dynamic can be seen in the gay male couple who would use cloning to obtain a genetically connected child, the lesbian couple who wish to have a child without any biological contribution from a male, or the sports fanatic who desperately wants a "son" with superior athletic talents. It has been reported that one of the reasons that the body of the Boston Red Sox power hitter Ted Williams was cryogenically frozen was the prospect that someday his DNA could become a valuable commodity. And serious discussions have been held about whether it would be a good idea to clone such sports stars as Michael Jordan and Tiger Woods.[44]

In contrast, opponents of CPC object to using cloning as a reproductive technology precisely because of the adverse impact it would very likely have on the cloned child, on the family and on society at large. As we have seen, the President's Council on Bioethics in its 2003 report made a powerful case against CPC based on safety considerations; rarely also has the moral case against it been so clearly explained. The council's objections centered on the problems that would face a cloned child, including:

Problems of Identity and Individuality

Cloning-to-produce-children," the council's report warned, "could create serious problems of identity and individuality. This would be especially true if it were used to produce multiple 'copies' of any single individual." (Theoretically, there is no limit to the number of SCNT clones of the same DNA donor that could be manufactured.)

Without speculating about the ramifications of creating bunches of almost genetically identical individuals, let's consider, as a thought experiment, the pressures that might be imposed on a single such person: for example, a child made from the DNA of a great musician with the intent that she follow in her genetic donor's footsteps. Rather than opening up wonderful opportunities for her, wouldn't her freedom actually be restricted by *biological constraints* forcing her in a predetermined direction chosen by her "parents"? What if she decided to become an engineer or a homemaker instead of a violinist? Would the disappointed cloners ask for their money back?

And what about the sad situation where the cloned child is intended to "replace" a dead genetic donor, as the bereaved parent wanted to do in Rick Weiss's *Washington Post* story? Wouldn't the genetically resurrected child—who, despite being a new and unique human individual— be haunted by her departed twin and be placed under tremendous pressure to emulate her, rather than forge ahead on her own path and in her own way? Or as the President's Council eloquently put it: "Living up to parental hopes and expectations is frequently a burden for children; it could be a far greater burden for a cloned individual. The shadow of the cloned child's 'original' might be hard for the child to escape, as would parental attitudes, that sought in the child's very existence to replicate, imitate, or replace the original."[45]

Concerns Regarding Manufacture

Cloning would remove procreation from the realm of the natural. Today, the only way a new human individual comes into being is when sperm and egg unite, whether in a woman's body or in a Petri dish. But cloning is by definition asexual: no sperm is involved. Likewise, the female contribution—the egg cell—is not making a chromosomal contribution to the new life; it is a mere receptacle, a favorable environment for the replication of the set of forty-six chromosomes provided by the DNA donor.[46]

In essence, therefore, cloning would treat procreation as a matter of manufacture—just as in *Brave New World*—resulting in "human products, brought into being in accordance with some pre-selected genetic pattern or design, and therefore in some sense, 'made to order,' by their producers or progenitors."[47] A possible consequence of reducing procreation to manufacture would be the "conditional acceptance of the next generation" by parents who decide not only "whether" to have a child but "*what kind* of child to have," and who see the procreative process as a "means of meeting specific ends." This matters, according to the council, because "human dignity is at stake."[48]

This transformation of childbearing from an intimate and personal human activity into one almost entirely in the hands of biotechnologists and technicians—and perhaps fully in their hands, should artificial wombs ever be perfected—would have significant social ramifications. "We would learn to receive the next generation less with gratitude and surprise," the council warned, "than with control and mastery." This in turn could lead to a profound coarsening of our society:

> One possible result would be the industrialization and commercialization of human reproduction. Manufactured objects [cloned children] become commodities in the marketplace, and their manufacture comes to be guided by market principles and financial concerns. When the "products" are human beings, the "market" could become a profoundly dehumanizing force.... The adoption of a market mentality in these matters could blind us to the deep moral character of bringing forth new life. Even were cloning children to be rare, the moral harms to a society that accepted it could be serious.[49]

Cloning Could Become a Tool for a New Eugenics

After it was discredited more than a half-century ago in the wake of Nazi eugenics, the theory that humankind can be improved by manipulating the gene pool has made a roaring comeback. Believers in the old eugenics accepted the propriety of harsh discrimination against the weakest and most vulnerable among us. Between 1920 and 1945, eugenics became a powerful social force aimed at cleansing the "genetically unfit" from the human family.

The evil of eugenics stemmed from its premise that because of certain traits, some people could be judged more worthy of life than others. Thus, it was not only acceptable but also desirable to manipulate the human genome so that the "fit" would proliferate and the so-called unfit dwindle away. Today, many pro-cloners have a similar perspective. They view human life as a malleable substance, subject to genetic enhancement or improvement. Nobel laureate James D. Watson, co-discoverer of the DNA double helix, foresees when genetic engineering and gene enhancement—technologies that would require cloning—could permit us to manufacture "perfect" humans.[50] The council worried that such a new eugenics "would seek to alter humanity, based upon arbitrary ideals of excellence," resulting in an "altering of human nature" and threatening a "post-human future."[51]

Troubled Family Relations

Consider for a moment the convoluted and mind-boggling relationships that might exist within a family into which a cloned baby was born:

> The usual clear designations of father and brother, mother and sister would be confounded. A mother could give birth to her own genetic twin, and a father could be genetically virtually identical to his son. The cloned child's relation to his or her grandparents would span one and two generations at once.... The crucial point is ... the presence of a unique, one-sided, and replicative and biological connection to only one progenitor. As a result, family relations involving cloning would differ from all existing family arrangements, including those formed through adoptions, or with the aid of IVF.[52]

To illustrate how surreal this could become, assume that my wife, Debra, gave birth to a cloned baby made from my DNA. Would I be the child's true father, or would his father actually be my late father—his "grandfather"—Wesley L. Smith? Would Debra be his real mother because she carried and gave birth to him, or would his mother be my mother, Leona? After all, the child's chromosomes and genes—passing from me to him through the SCNT process—would have been created originally out of the blending of my parents' chromosomes. I get my height and large frame from my father's side of the family, my nose, hair color and voice timbre from my mother's. Probably, so would he. On the other hand, there would be virtually nothing of Debra's side of the family in him, other than mitochondrial DNA. If we used another woman's egg cell, there would be nothing of Debra at all.

And imagine if we got a child by cloning my late father: My "son's" real father would actually be my grandfather, and I would be raising my father's near identical twin! Or what about the possibility that a woman will one day give birth to a clone of herself after she has already borne children by natural conception? Not only would she have given birth to her own twin sister, but the "child that is born will become the genetic mother of her older brothers and sisters."[53]

Effects on Society

Human cloning has radical implications for our society as a whole. Just as society was profoundly affected by moral issues such as human slavery, women's suffrage and the struggle for civil rights, so too would our society be profoundly changed should cloning ever be deemed a legal and/or legitimate method of procreation.

The President's Council on Bioethics does not shrink from this truth. "The impact of human cloning on society at large," its report states, "may be the least appreciated, but among the most important, factors to consider" in contemplating whether society should permit it. The council noted that if cloning-to-produce-children (CPC) were permitted, society would "become an accomplice" to it and in effect could be said to "engage in it," thereby becoming complicit in all the negative consequences that might ensue. At stake is "what kind of a society we wish to be." The council predicts,

> Cloning-to-produce-children could distort the way we raise and view children, by carrying to full expression many regrettable tendencies already present in our culture. We are already liable to regard children largely as vehicles for our own fulfillment and ambitions. The impulse to create "designer children" [that human cloning could facilitate] is present today—as temptation and social practice. The notion of life as a gift, mysterious and limited, is under siege. Cloning-to-produce-children would carry these tendencies and temptations to an extreme expression. It advances the notion that the child is but an object of our sovereign mastery.[54]

Ample evidence that the council's warning is not exaggerated can be found in the writings of cloning supporters. Leading the way before the cloning of mammals was widely considered possible, Joseph Fletcher urged that we cast aside natural procreation in favor of biotechnological manipulations—forgetting or ignoring thousands of years of experience and wisdom about the importance of the biological family to the survival of societies and to individual well-being. In Fletcher's jargon:

> Parental (and kin) relationships need to be reconceptualized. They cannot any more be based on blood or wombs or even genes. Parenthood will have to be understood non-biologically or, to be specific, morally. . . . Parental love [has to become] truly impersonal; no longer can it be merely germinal, somatic, or physiological—and certainly not merely genital. An

authentic parental bond is established morally, by care and concern, not by some simple physicalist doctrine.[55]

Even the foreseeable consequences that would flow from CPC would be radical and far-reaching—and there would very likely be other shattering consequences that we can't, as yet, foresee. Thus, there is much more at stake here than the protection of cloned children; the issues strike at the very concept of what it means to be human.

At this turning point in history, outlawing CPC—at least until it is "safe"—is utterly uncontroversial. The same cannot be said about prohibiting cloning-for-biomedical-research. Many current opponents of CPC have few qualms about CBR, arguing that we could mine cloned embryos for their stem cells without running headlong into the dangers we have been discussing. I believe they are wrong; here the hackneyed "in for an inch, in for a mile" could hardly be more pertinent.

CHAPTER 4

The Foot in the Door to a Brave New World

· ·

C ONSIDER THIS BRIEF TIMELINE. In 1997, Dolly was born, making human cloning a possibility. In 1998, the first human embryonic stem cell lines were established. In that same year, scientists at Japan's Kinki University cloned eight identical calves using cells taken from a single adult cow, raising the prospect that humans could be similarly mass-produced. In 2002, genomic researchers published the draft version of the complete map of the human genome, identifying the location of approximately 30,000 human genes—an essential preliminary to genetic engineering, among other things.[1] In 2003, a scientist from the People's Republic of China announced that she had successfully created human cloned embryos via SCNT using rabbit instead of human eggs.[2] Finally, in 2004, a researcher in South Korea published data demonstrating that he had succeeded in creating thirty human cloned embryos, from which he had succeeded in isolating one embryonic stem cell line.[3] In other words, in seven short years we have gone from a cloned sheep to cloned nascent humans.

This is a pivotal moment. The United States and much of the world is engaged in a protracted political and moral debate about how to best come to grips with the prospect of human cloning. One side—the side I support—believes that the only sure way to control human cloning is to prohibit it altogether. The other side is less forthright. Knowing that the prospect of human cloning upsets most people, advocates assure us that they are not at all interested in bringing forth cloned babies. But they go on to argue that human cloning-for-biomedical-research (CBR) should be explicitly legalized; otherwise, they darkly prognosticate, society will never reap the fruits of regenerative medicine. Carl B. Feldbaum, president of the Biotechnology Industry Organization (BIO), a powerful lobbying arm for biotech, explained the industry's opposition to a proposed two-year moratorium against human cloning in a letter to Senator Orrin Hatch (R-UT), who is a proponent of CBR:

> SCNT research is vital because it may harness the potential of stem cells and move information from the lab into the doctor's office as cures and treatments. Many diseases and disabilities are caused by disrepair and damage to cells and tissue. In conjunction with stem cell research, cloning technology could be used to develop products such as replacement cells and tissue.... Congress must not pass a moratorium that would halt the important research needed to treat deadly and debilitating diseases afflicting millions of Americans.[4]

The United States, which should lead the world in determining guidelines for biotechnology, finds itself politically paralyzed. It isn't that the American people don't want to enact effective national controls over human cloning—we do. To date, total human SCNT cloning bans have twice passed the U.S. House of Representatives in lopsided, bipartisan votes. Senators Sam Brownback (R-KS) and Mary Landrieu (D-LA) are pushing a similar bill in the Senate, and President George W. Bush has promised to sign the legislation into law should it reach his desk. But as these words are written, a filibuster against the ban has stalled its passage in the United States Senate. As a result, the U.S. still has no federal law that regulates or constrains experiments in human cloning.

Phony Bans

A competing bill to Brownback/Landrieu is S. 303, authored by Senators Orrin Hatch (R-UT) and Dianne Feinstein (D-CA). This measure, a purported "compromise" between an outright ban and unlimited legality, is typical of the approaches being trotted out around the world by cloning supporters. The senators misnamed their bill the "Human Cloning Ban and Stem Cell Protection Act of 2003." In reality, Hatch/Feinstein would not have outlawed human SCNT cloning at all; on the contrary, it would explicitly have legalized it, while not allowing human cloned embryos to become fully developed babies. This is why I shall hereafter refer to proposals of this sort not as "partial bans," as they are sometimes called, but more accurately as "phony" cloning bans: they don't ban human cloning at all. (In point of fact, theirs is a truly radical proposition. Legalizing human SCNT would permit biotechnologists to create human life intentionally through cloning and *legally require* that it be destroyed.)

So how did Senators Hatch and Feinstein justify putting the term "cloning ban" into the name of their legislation? Instead of defining cloning accurately and scientifically as the SCNT procedure itself, the bill redefines the term to mean "implanting or attempting to implant the product of nuclear transplantation [i.e., the cloned embryo] into a uterus or the functional equivalent of a uterus."[5] In other words, rather than focusing on the actual act of cloning, Hatch/Feinstein changed the term's meaning to refer to what is done with the cloned embryo *after it has been created.*

Hatch/Feinstein also redefined human cloning, when not conducted to produce children, as merely a form of stem cell research. Thus, a cloned human embryo becomes "an unfertilized blastocyst," which the proposal calls an "intact cellular structure that is the product of nuclear transplantation." But as we have already noted, the product of human nuclear transplantation is a cloned human embryo—far more significant biologically and profound morally than a mere cellular structure. It takes a week of development and growth for an embryo to reach the blastocyst stage when its cells can be extracted to create a stem cell line. The bill changes this basic biological definition by simply calling a cloned embryo a blastocyst both before and after it attains that actual stage of development.

As a political water carrier for Big Biotech and the science establishment that seek a license to engage in CBR, Senator Feinstein has frequently muddled the stem cell debate in this way. Indeed, she once took to the floor of the U.S. Senate to argue that a cloned human embryo, from which embryonic stem cells would be extracted, was not really an embryo at all, but merely an "unfertilized egg":

> The beauty of our legislation is that it would allow this most promising form of stem cell research, somatic cell nuclear transplantation, to be conducted on a human egg for up to fourteen days only, under strict standards of Federal regulation.... The reason for this fourteen days is to limit any research before the so-called primitive streak [when neural cells begin to appear] can take over that egg. This stem cell research can only take

place on an unfertilized egg. . . . An unfertilized egg is not capable of becoming a human being. Therefore we limit stem cell research to unfertilized eggs.[6]

This is nonsensical on its face. An unfertilized egg doesn't develop embryonic stem cells; embryos do.[7] An unfertilized egg cannot develop a primitive streak on its own. An unfertilized egg is merely a cell, a "gamete" in the scientific lexicon. In contrast, the product of SCNT would be a cloned human embryo, which if allowed to develop into the second week would develop a primitive streak.

Such obfuscation is part of an overarching strategy. This and similar proposals are worded so as to confuse. While most people oppose human cloning, whether for reproductive or experimental purposes, they tend to support ESC research if the embryo to be destroyed is left over from IVF procedures and will be discarded anyway. Playing on these scruples, pro-cloners simply pass legislation that redefines cloning-for-biomedical-research as a form of ESC research.

When deceptive terminology is employed to make it appear that only "cells" are created in a "therapeutic cloning" procedure, public support tends to grow. Remarkably, even genetic scientists can be swayed by such verbal subterfuge: they tend to express different attitudes toward "therapeutic cloning" depending on how a polling question is asked. Thus *Genetic Engineering News* published the results of a survey that asked American and international biotech researchers about their moral attitudes toward human cloning and ESC research. A total of 1,229 scientists and researchers from the United States responded to the surveys, and 408 from abroad. The results: 92 percent of U.S. and 85 percent of international scientists support "therapeutic cloning of human cells for replacement tissue."[8] *Business Wire* concluded that "a majority of international scientists favor . . . the therapeutic cloning of cells."[9]

No surprise here. Yet when the question about the same procedure was posed in a different way, support for therapeutic cloning among the surveyed scientists plummeted. Only 34 percent of U.S. and 19 percent of foreign survey respondents supported "human cloning research without implantation of the cloned embryo in a human uterus."[10] But this is exactly the same procedure as "therapeutic cloning of cells," only described more accurately. Moreover, a whopping 73 percent of U.S. and 78 percent of international scientists *"believe the creation of human embryos specifically for research purposes is ethically unacceptable.*"[11] But this is precisely what therapeutic cloning is intended to do!

If even the experts are confused, the public is more so, which explains why Big Biotech's lobby refuses to use accurate definitions and make lucid arguments. It wants to win and isn't about to let facts get in the way.

Phony Bans on "Reproductive Cloning" Won't Work

A law prohibiting cloned embryos from being brought to birth but not also outlawing CBR would actually end up promoting such births. The Ethics Committee of the American Society for Reproductive Medicine admits that "If undertaken, the development of SCNT for ... therapeutic purposes, in which embryos are not transferred for pregnancy, is likely to produce knowledge that could be used to achieve reproductive SCNT."[12] And Woo Suk Hwang, the South Korean researcher who first successfully created a cloned human blastocyst for the extraction of embryonic stem cells, told journalists, "I do not rule out the possibility that our findings can be used in reproductive cloning by some rogue scientists."[13]

In fact, some unprincipled scientists are already seeking to perfect human SCNT for purposes of reproduction (CPC), as Panayiotis M. Zavos makes clear in his article "Human Reproductive Cloning: The Time Is Near." In addition to creating human/animal hybrid cloned embryos, Zavos announced that his "team of scientific and medical experts" had "created the first cloned embryo for reproductive purposes." They were able to develop the embryo, which he contended was manufactured with a human egg, for four days post-SCNT, until it reached "the 8–10 cell stage" and showed "a rate of development equivalent to that of normal IVF embryos." The embryo was frozen for future molecular analysis, toward the end of eventually implanting such an embryo in a woman's body.[14] (This was later reportedly done, resulting in miscarriage.) This is precisely the kind of experiment that might make it possible someday to create a cloned human baby, and it is an experiment that a phony ban would *explicitly sanction.*

Dr. Zavos and other bio-anarchists such as Lee M. Silver and Gregory Pence are not the only ones who would support CPC should researchers ever overcome the existing safety problems. For example, the Clinton administration's National Bioethics Advisory Commission (NBAC), asked by the president to consider the future of human cloning in the immediate wake of Dolly's birth, recommended a ban-cloning-for-now-but-probably-allow-it-later approach, urging that Congress enact only a three-to-five-year prohibition against attempting to "create a child through somatic cell nuclear transfer cloning." When the moratorium elapses, NBAC suggested that an "appropriate oversight body" be created—presumably made up of bioethicists like themselves—to "report on the current status of somatic cell nuclear transfer technology and on the ethical and social issues that its potential use to create human beings would raise in the light of public understandings at that time."[15] In other words, if it ever became safe, and if the people of the United States could be sufficiently desensitized to the moral wrongness of human cloning, the NBAC foresaw the distinct possibility of permitting CPC.

The Ethics Committee of the American Society for Reproductive Medicine adopted a similar view in 2000. The committee agreed that "reproductive SCNT does not meet standards of ethical acceptability" at this time. However, "It does not necessarily follow that the procedure should be foreclosed permanently." Moreover, a "moratorium on reproductive SCNT" would not "remove the need to study more carefully the ethical implications of cloning, especially for infertile couples," meaning that should biotechnologists engaging in Hatch/Feinstein-authorized cloning research learn how to make CPC "safe," the chances seem good that the committee would view cloning in the same light as IVF.[16]

The National Academy of Sciences (NAS) has been the most notable scientific organization weighing in forcefully in support of the phony-ban approach.[17] "Because medical and scientific findings indicate that cloning procedures are currently not safe for humans," the NAS states in *Scientific and Medical Aspects of Reproductive Cloning*, "cloning of a human through the use of nuclear transplantation technology is not now appropriate."[18] (The key word in this sentence is "now.") Like the NBAC before it, the NAS did not propose that reproductive cloning be permanently outlawed, but rather that it be "reviewed within five years," and that any prohibition be reconsidered if "a new scientific and medical review indicates that the procedures are likely to be safe and effective," and if "a broad national dialogue on the social, religious, and ethical issues suggests that a reconsideration of the ban is warranted."[19]

As a matter of moral principle, the majority of bioethicists and biotechnologists probably do not object to the creation of cloned human children. Some authors view the right to reproduce as so absolute that it must permit access to virtually any biotechnology that persons desiring to have children wish to employ, assuming its relative safety.[20] Indeed, many advocates see cloning as a splendid way for infertile or homosexual couples to have biologically related children.[21] Others, including Dolly's maker, Ian Wilmut, writing with University of Pennsylvania bioethicist Glenn McGee, would treat cloning as states now treat requests to adopt children, that is, by establishing a bureaucracy to review and approve reproductive cloning requests.[22] Presumably, given that the procedure is hazardous, these would at first be refused. But if safe CPC ever became feasible, the door already ajar could in principle be easily opened wide. In later advocacy, Wilmut went even further, urging that prospective parents be allowed to have a cloned baby as a way to avoid genetic disease.[23]

In the aftermath of the Raelian hoax, an article written by *Washington Post* columnist Richard Cohen demonstrated how desensitized some have already become to the prospect of bringing cloned children into the world. Once cloning became safe, Cohen wrote, it "would just be another form of reproduction." Cohen proposed divorcing morality and ethics from the cloning controversy altogether, declaring that "terms like 'ethical' or 'human dignity' simply cloud the debate."[24]

So this is the bottom line: It is indisputable that researching human SCNT "can prepare the way scientifically and technically for efforts at reproductive cloning."[25] Some, like Cohen, Zavos and Rael, among others, are already preparing the moral ground to permit human cloning to proceed unhindered when the time is ripe. Should scientists learn how to manufacture cloned embryos without defects, it doesn't take a psychic to predict what would follow under the auspices of a law of the Hatch/Feinstein variety:

Bring On the Judges. Recent history suggests that courts now serve as mini-legislatures in creating social policy, a phenomenon that would undoubtedly prevail in the cloning controversy as well. Should it be deemed relatively safe to bring cloned babies to birth, several infertile and/or homosexual couples will undoubtedly file a lawsuit claiming that the ban on reproductive cloning violates their "fundamental right" to procreate. Considering the importance that recent court decisions have placed on the right to reproduce, it's quite conceivable that any ban would be declared unconstitutional. (The *National Law Journal* has already reported that preparations are under way within the legal community for this eventuality.)[26]

Bring On the Talk Show Hosts. Concurrently with the lawsuits, look for couples to appear on *Oprah* urging an end to the ban on CPC so that they can fulfill their desperate desire to have biologically related children. With human cloning now theoretically safe, they and their supporters in the biotech industry will argue, cloning should be treated as just another reproductive technology akin to IVF.

Bring On the Mad Scientists. Once human cloned blastocysts can be reliably manufactured, some fame-seeking scientist, perhaps working offshore, will implant one for a woman who wants to go down in the *Guinness Book of World Records* as the first birth mother of a human clone. Since no one will urge that she be forced to undergo an abortion, at that point the birth of a cloned baby would be virtually unstoppable.

Clearly, if human SCNT is ever legalized, and hence legitimated, sooner or later the likely result would be the birth of cloned human children. So if preventing cloned children is our goal, then illegalizing all human SCNT is the only reasonable course.

Therapeutic Cloning Is More Mirage Than Reality

The most potent arrow in the pro-cloning quiver is the promise that "therapeutic cloning" will become a panacea for some of humankind's most intractable degenerative diseases. This is how BIO, Big Biotech's lobbying association, touted therapeutic cloning as a treatment for degenerative diseases in 2003:

> Suppose a middle-aged man suffers a serious heart attach while hiking in a remote part of a National Park. By the time he reaches the hospital, only a third of his heart is still working, and it is unlikely he will be able to return to his formally active life. He provides scientists a small sample of skin cells. Technicians remove the genetic material from the cells and inject it into donated human eggs from which the chromosomes have been removed. These altered eggs will yield stem cells that are able to form heart muscle cells. Since they are a perfect genetic match for the patient, these cells can be transplanted into his heart without causing his

immune system to reject them. They grow and replace the cells lost during the heart attack returning him to health and strength.[27]

Let's ignore for the moment that in early human trials, heart patients have already received identical treatments, but using their own blood or bone marrow stem cells.[28] And never mind that rejection would definitely not be a problem with adult stem cells, since they are the patient's own tissues, whereas rejection could be a problem with the heart patient's cloned cells owing to the presence of mitochondrial DNA from the donor's egg. And set aside the fact—contrary to the implications of the BIO scenario—that developing a stem cell line for the needy patient could take months, perhaps longer, to isolate and characterize. The hard truth is that even if therapeutic cloning of the kind envisioned could be performed, it is still so unrealistic that pursuing cloning in the search of regenerative cures is akin to chasing after a mirage.

Remember that the sales pitch on behalf of legalization states or strongly implies that once developed, cloning as a medical treatment would be available to many, if not most patients afflicted with degenerative illnesses or injuries. This is quite a large number of people. According to the National Academy of Sciences, more than 130 million Americans suffer from degenerative conditions such as cardiovascular disease, diabetes, Alzheimer's disease, Parkinson's disease, severe burns, spinal cord injuries, diabetes and cancers. And while not everyone with these diseases could benefit from "stem-cell-based therapies," the NAS explicitly stated that "stem-cell research could help millions of Americans."[29]

More probably, patients who hope to be treated with therapeutic cloning are fated to be sorely disappointed. Why? Because there will simply never be enough human egg cells available to make all the cloned embryos required for the tens of millions of patients who would be clamoring for treatment.

David A. Prentice, a biologist, ASC researcher and former professor of life sciences at Indiana State University, is one of the world's most knowledgeable scientists when it comes to the biotechnological issues involving human cloning. Traveling widely, lecturing before august committees and legislative bodies, he strongly endorses outlawing human SCNT. After repeatedly listening to biotech boosters hold out the prospect of therapeutic cloning as a panacea for millions of patients, Prentice decided to crunch some numbers to see how many human eggs would be necessary for it to become widely available in American clinics.

In order to keep the numbers manageable, Prentice analyzed the number of eggs that would be required just to offer treatment to the 16 million Americans listed as diabetics by the NAS at the time he made his estimate. He assumed a 20 percent success rate in developing cloned embryos to the blastocyst stage. (Prentice based this number on published reports in peer-reviewed journals on the efficiency of obtaining blastocysts through animal cloning.)

Next, he assumed that ESC lines would be successfully derived from 10 percent of these blastocysts. This figure is also fair—it took 36 embryos for James Thomson of the University of Wisconsin to create five human ESC lines, a 13.8 percent success rate, while the Jones Institute used 110 embryos to get only three stem cell lines, a 2.7 percent success rate. In March 2004, Harvard University scientists announced that they had created 17 viable embryonic stem cell lines and that the effort required 344 embryos, about a 5 percent success rate.[30] And bear in mind that these were natural embryos created through fertilization, a process that doesn't pose the gene imprinting problem found in mammalian embryos cloned with adult somatic cells.

Using these numbers, Prentice discovered that it would take 800 million eggs to provide therapies for the 16 million diabetics in the U.S.A.[31] Recall that eggs are extracted from women of childbearing age by means of hyperovulation, whereby a woman's body is stimulated with hormones to release between seven and ten mature egg cells in her monthly cycle rather than the usual single egg. Assuming a generous estimate of ten eggs harvested from each procedure, Prentice determined that it would require about 80 million egg donations from women of childbearing age just to treat America's diabetics. With this arithmetic— there aren't that many women of childbearing age in the United States, obviously—he found what is probably an insurmountable barrier to therapeutic cloning ever becoming a widely available medical therapy.

Many were skeptical when Prentice first revealed the utter impracticality of therapeutic cloning. But now, even some of the most enthusiastic boosters of CBR have concluded that it "will be too expensive and cumbersome for regular clinical use."[32] For example, James Thomson, the researcher who first isolated human ESCs, has written, "The poor availability of human oocytes [eggs], the low efficiency of the nuclear transfer procedure, and the long population-doubling time of human ES cells make it difficult to envision this [therapeutic cloning] becoming a routine clinical procedure. "[33]

In "Therapeutic Cloning in the Mouse," an article published in 2003 by the National Academy of Sciences, Peter Mombaerts of Rockefeller University gives further credence to the egg-dearth thesis. Mombaerts, a supporter of cloning research, has spent years trying to develop therapeutic cloning techniques in mice. It has apparently been tough going, and his report at a scientific colloquium was gloomy: "The efficiency [of SCNT], or perhaps better, the lack of efficiency thereof, is remarkably consistent."[34]

"Lack of efficiency" doesn't even begin to describe the problem. Prentice told me that he estimated that each therapeutic cloning attempt would require about fifty eggs to obtain one viable ESC line, meaning of course that approximately fifty eggs would be needed per patient. This may be overly optimistic. Based on his experiences with mouse cloning, Mombaerts suggested that it would take approximately one hundred eggs to obtain just one ESC line per patient.

Nor did Mombaerts expect these daunting numbers to improve as biotechnologists gained experience, given that "the efficiency of nuclear transfer has not increased over the years in any of the mammalian species cloned."[35] Making matters even more difficult, researchers have had tremendous difficulty accomplishing human and primate SCNT because of chromosomal problems that do not exist in the cloning of mice. Hence deriving cloned ESC lines from humans could well be significantly less efficient than it has been to date with animals.[36] (Indeed, it took South Korean scientist Woo Suk Hwang 242 eggs to create the first human cloned ESC line.)[37]

This is big news. If it takes one hundred or more tries to make a single human cloned embryonic stem cell line, therapeutic cloning is doomed as a practical medical treatment because of an intractable insufficiency of resources. Since there are perhaps more than 100 million Americans who might benefit from tissue regeneration therapy, it would require at least 10 billion eggs to make cloned ESCs available for all of them.

Even if the (theoretical) benefits of therapeutic cloning were restricted by a draconian system of rationing to, say, the sickest 100,000 patients in the United States, biotechnologists would still require access to at least 10 million egg cells. Under such a regime, approximately one million women of childbearing age would be required to submit to egg extraction.

Therapeutic cloning would also be very expensive. At present, women sell their eggs for use in fertility treatments, according to Mombaerts, for $1000–$2000 apiece. He concludes that the cost per patient for the eggs alone—which does not include the expense associated with doctors, hospitals, laboratories, etc.—would run in the neighborhood of $200,000, "a prohibitively high sum" that Mombaerts expects "will impede the widespread application of this technology in its present form."[38]

According to *Nature,* many biotechnologists have begun to see the writing on the wall. The prestigious science journal's chief news and features editor wrote:

> Enthusiasm for therapeutic cloning was initially high. So, to the casual observer, it may come as a surprise that *many experts do not now expect therapeutic cloning to have a large clinical impact....* Aside from problems with the supply of human egg cells, and ethical objections to any therapy that requires the destruction of human embryos, many researchers have come to doubt whether therapeutic cloning will ever be efficient enough to be commercially viable. "It would be astronomically expensive," says James Thomson of the University of Wisconsin in Madison, who led the team that first isolated ES cells from human blastocysts.[39]

And if the price tag of therapeutic cloning would be exorbitant today, imagine how expensive it would become should the demand for egg cells skyrocket as a result of the desire of millions of patients to undergo treatments. Basic economic theory predicts that a soaring demand in the face of a very limited supply results in price hyperinflation. In late 2002, a report out of New Jersey described how egg cell prices had tripled in only four years to $7,500 per egg harvesting procedure because of a "fiercely competitive market to supply eggs to patients" desiring fertility treatments.[40]

All this adds up to the fact that *if* therapeutic cloning could ever be developed—still a gargantuan if—it would be prohibitively expensive for government funders, health insurance companies and everyone but the richest people. So it would have to be either rigorously rationed to a select few, perhaps by a lottery system, or available only to the hyperwealthy.

But there is yet another element we should consider: the danger that widespread human cloning would lead to the exploitation of women as so many farms for the production of egg cells. There is a distinct possibility that biotechnologists, hungry for the egg cells needed in cloning, would turn to desperately poor and uneducated women in undeveloped nations: many would undoubtedly sell their egg cells at bargain-basement prices.

This is not just a speculative concern. It is already a fact that the shortage of human organs needed for transplantation has resulted in an international black market that exploits and victimizes the most impoverished. The British Broadcasting Corporation reported, for example, that poor people in India might sell a kidney for as little as $700.[41] Desperate people in Turkey may also feel they have no other recourse but to become organ peddlers; it was reported recently that a father sold one of his kidneys to pay for an operation for his child.[42]

The world's poorest women would no doubt fall prey to the insatiable demand created by the prospect of therapeutic cloning. Given that many of these destitute egg cell suppliers would not have ready access to medical care, it's worth considering the significant health risks associated with the process of egg extraction.

Egg cell procurement involves the use of strong hormones to "hyperstimulate" the subject's ovaries so that she releases multiple mature eggs during her monthly cycle. The eggs are then surgically removed—generally with a needle inserted through the vaginal wall. Significant complications can result requiring ready access to good medical treatment. For example, one out of one hundred women experiences ovarian swelling that may enlarge her ovaries to the size of a grapefruit, a condition that often needs hospitalization. Some even suffer ovarian rupture.[43] Other potential health hazards include infection, pulmonary and vascular complications, including the life-threatening acute respiratory distress syndrome, blood clots and future cancers. In rare cases, hyperovulation results in death. These and other risks are so serious that some ethicists have called for a legal ban prohibiting women from donating eggs to strangers.[44]

For women under a doctor's care, the dangers associated with egg cell procurement are usually manageable. But that would often not be true for women living in impoverished countries—or perhaps for illegal aliens living in wealthy ones—who generally do not have ready access to medical services or who may be afraid to seek a doctor's help. We must ask ourselves: Would it be right to willingly endanger destitute women in order to obtain their egg cells for therapeutic cloning?

Is there any way out of the human egg deficit? Cloning researcher Peter Mombaerts, whose work I discussed earlier, suggested two potential sources of egg cells other than women, to which I add a horrific third:

Animal eggs. These are more readily available than human eggs, which would reduce the price of therapeutic cloning considerably. But using animal eggs in human SCNT would entail the sanctioning of hybrid human/animal clones.

Some experimenters have already created human/animal hybrid cloned embryos. Panayiotis M. Zavos claimed to have made human/cow hybrids as part of learning how to engage in CPC. Advanced Cell Technology boasted several years ago about having conducted similar research. And in China a research team led by Huizhen Sheng claimed to be creating cloned human embryos using rabbit eggs. "There is a possibility," she told *Nature* in 2003, "that if it were proved to be safe enough for clinical use, it could provide a solution to human egg shortages."[45]

If creating human/animal hybrids ever became a widespread practice, would the American people stand for such a violation of nature's laws? Mombaerts worries that such an "Island of Doctor Moreau" undertaking would undoubtedly generate heated controversy: "The idea of generating embryos with mixed human/animal properties, even transiently, is offensive to many people."[46]

Even from a purely practical standpoint, the idea of using stem cells derived from animal eggs is highly problematic. All of the cells and tissues derived from human/animal hybrids would inevitably contain animal mitochondrial DNA; how this would affect patients injected with such cells is not known and would probably take a long time to find out. But there is a distinct possibility that the body would reject them, or that they could cause animal-derived illnesses.[47]

Transforming embryonic stem cells taken from fertilized embryos into fully formed human eggs. Since in theory ESCs can transform into any tissue, some believe they could be induced to transform into a bounteous supply of human eggs. Mouse stem cells have been generated into eggs, but it will take many years to determine whether this feat can be accomplished using human ESCs. Even if researchers someday succeed, that does not mean the egg cells so derived would be ready for immediate use in cloning. First, researchers would have to conduct extensive testing to determine whether such egg cells would be safe for use in therapeutic cloning procedures. Second, the nascent eggs would have to be matured for use in cloning, a process that is also still in the experimental stages. And even if researchers over many years successfully surmounted these hurdles, considering the hundreds of millions of eggs that would have to be produced in this way for therapeutic cloning to become widely available, morphing eggs out of ESCs hardly seems a plausible answer to the egg cell dearth conundrum.

We could use the ovaries of aborted female fetuses to develop egg cells. I know this is revolting, but Dutch and Israeli researchers are already experimenting with maintaining the ovaries of second- and third-trimester aborted female fetuses to determine whether they could become a source of eggs for use in infertility medical treatments.[48] Not only does this macabre research open the strange possibility that an aborted baby girl could become a mother, but the procedure, if perfected, could result in aborted female fetuses becoming a commodity. As if that weren't troubling enough, the abortions of these fetuses might have to be done in such a way that their ovaries would be left intact, providing utilitarian impetus for the odious procedure called intact dilation and extraction (D&X), popularly known as partial-birth abortion.

Human Cloning Research Would Never Stop Expanding

The limits that a "phony ban" of the Hatch/Feinstein type proposes—it would legitimate human SCNT but ban implantation of the resulting cloned embryo into the womb—would be unlikely to satisfy cloning boosters' appetites for very long. Rather, should it become acceptable to maintain cloned embryos through the blastocyst stage, biotechnologists would soon lobby for an even broader right to exploit cloned human life—once again justifying their request by promising miracle medical cures.

As we have already seen, this would probably include a political assault on the prohibition against cloning-to-produce-children, once it was declared "safe." But even before seeking legal sanction for the creation of cloned human babies, chances are that Big Biotech, hungry for profits and infusions of new venture capital, would urge that they be permitted to implant cloned embryos, have them gestated into fetuses, and then aborted prior to birth for use in research or extraction of tissues for medical treatments.

Is this paranoid? I wish it were so. Expanding the scope of cloning research beyond the earliest embryonic stages already heads the agendas of Big Biotech and the scientific establishment. Stanford University law professor Henry T. Greely is the director of Stanford's Center for Law and Biosciences. As a member of the California Advisory Committee on Human Cloning, he supported a recommendation that California legalize human SCNT and provide state funding for the research, with the proviso that the embryos be destroyed after the "development of the primitive streak" at about fourteen days.[49] The State of California subsequently passed such a law, legalizing human SCNT and authorizing state funding to help pay for the research.

It soon became clear that the fourteen-day limit was merely an opening gambit. A few months after the publication of the California Advisory Committee's report, Greely participated in a "neuroethics" symposium in San Francisco.[50] According to the official published transcript, Greely freely admitted under questioning that the fourteen-day/primitive streak research limitation was merely a way station on the road to permitting a far more extensive cloning research license.

> *Question:* I wonder why, in your state commission's report on cloning, you folks recommended that therapeutic cloning be permitted, but only up to the formation at fourteen days of the primitive streak. That's the only part of the report I was disappointed in, because you went no further than all of these other commissions. Yet it seems to me that at fourteen days, you're still very far from an entity that is capable of consciousness, and by stopping there we'll likely be unable to investigate why embryos don't implant, for example, or how brain abnormalities like anencephaly develop. Can you comment on that?
>
> *Greely:*... That qualification was driven, I think it's fair to say, largely by two things: a very strong desire to have a unanimous report and the fact that it was fairly straightforward albeit a very conservative place to stop, at least for now, based on our current state of knowledge. Before cells begin to differentiate in their functions, it seems very hard for anyone to argue that there is the remotest chance that sentience exists in that small ball of cells. Even past the development of the primitive streak ... my own guess is that any neurological functioning will not come for many days and weeks. But fourteen days was a good, easy, clear stopping point *for now,* based on our current understanding. *We did not mean that fourteen days would always be the limit; that limit could be changed in the future based on new understandings that would likely come from neuroscience.*[51]

It's important to note that an unfrozen embryo can be maintained outside a woman's body only for about seven to at most fourteen days. Thus, implicit in the question and Greely's answer is that researchers want to be allowed to implant and use in research cloned embryos in real or artificial wombs for extended periods. Indeed, Greely's only apparent stopping point would be the emergence in the cloned human life of sentience or consciousness, which some bioethicists claim does not occur until some time after birth.[52]

Legislation already introduced in a few states offers further proof that Big Biotech's ambitions extend far beyond the widely touted fourteen-day research limit. Under these proposals, cloners would be allowed legally to implant cloned human embryos into wombs and gestate them through the ninth month; they would only be required to kill them *at the very point of birth*. Astoundingly, in 2004 New Jersey enacted just such a law.

The New Jersey Senate bill (S. 1909) was a sneaky piece of legislation.[53] According to the official synopsis, it was purportedly about "permitting human stem cell research in New Jersey." Its advocates publicly proclaimed that the law would allow ESC research from embryos left over from IVF procedures. But ESC research was already legal under federal law, and no proposals had been put forward in either the House of Representatives or the Senate to outlaw it—hence if that were all S. 1909 sought to permit, it would have been redundant.

But there was a far more disturbing point to the legislation. Hidden within the text of the amended bill lurked a shockingly radical agenda:

- The law explicitly authorizes the manufacture of human cloned embryos via the SCNT cloning procedure.[54]
- Unlike the Hatch/Feinstein measure, the law does *not* proscribe the implantation of cloned embryos into a woman's womb. Indeed, it is silent on this point. This is important because if an act is not illegal, by definition it is legal.
- The legislation criminalizes the "cloning of a human being" as a "crime of the first degree."

This sounds like a tough criminal sanction. But the devil is in the details. The key to understanding the dangerous depth and scope of S. 1909 is found in the law's definition of the term "human being":

As used in this section, "cloning a human being," means the replication of a human individual by cultivating a cell with genetic material *through the egg, embryo, and fetal and newborn stages into a new human individual.*[55]

Because of this wording, it is now legal in New Jersey to create a human cloned embryo via SCNT, implant it in a willing woman's uterus, and gestate it through the ninth month. As long as the fetus is killed before or at the moment of birth—that is, before entering the "newborn stages" of life—the law will not have been broken.

Because gestation and development of cloned fetuses up to the point of actual birth does not violate any law, four members of the President's Council on Bioethics criticized the law prior to its passage as abhorrent, saying that for "New Jersey (Senate Bill 1909/Assembly Bill 2840) to authorize human cloning and the harvesting and use of body parts of cloned humans in the embryonic and fetal stages of development" threatens to make the state "a haven for unethical medical practices, including the macabre practice of human fetal farming."[56]

But New Jersey is not alone. Texas state senator Eliot Shapleigh (D-El Paso) introduced an almost identical proposal in the Texas State Senate in 2003.[57] Like the New Jersey bill, Texas's SB 1034 purported to authorize the already legal ESC research. Like New Jersey's S. 1909, it also would have authorized the creation of human cloned embryos through SCNT and would not have outlawed their implantation. Moreover, just as in New Jersey, the Texas bill would have made it a crime to engage in "human cloning." But once again, human cloning was defined as "the replication of a human individual by cultivating a cell with the individual's genetic material through the egg, embryo, fetal, and newborn stages into a new human individual."[58] In other words, the Texas bill amounted to a second attempt to authorize cloned fetal farming. (The bill died when no action was taken on SB 1034 before the end of the 78th legislative session.)

Delaware also got into the cloning licensing act in 2003. The Delaware bill (SB 55), disingenuously entitled the "Cloning Prohibition and Research Protection Bill," actually would have *explicitly permitted* human cloning, implantation, and gestation—so long as the purpose of these actions was not to bring a cloned baby to birth.[59]

Another pro-cloning bill, introduced in 2003 in Maryland, while less explicit in its agenda than the New Jersey and Texas measures, might have led to the same radical license. The bill, HB 482, entitled "An Act Concerning Stem Cell Research—Donation of Certain Tissue for Research Purposes," would have permitted virtually any experiment into human cloning that researchers might devise. It would have explicitly legalized human SCNT. The bill was also silent about implanting cloned embryos into wombs for gestation, implying that it would be legal to do so. Moreover, no restrictions were placed on the types of experiments that could be conducted: the bill only mandated that "full consideration be given to the ethical and medical implications of this research" and required an internal review by university or corporate institutional review boards (IRBs).[60] In other words, it would be up to these private panels to decide what could or could not be done without any guidance from the legislation. Considering that the IRBs would very likely be made up of pro-cloning/pro-biotech industry members, in theory there would be no constraints on the cloning experiments that could be conducted had this bill become law.

One law couched in Brave New World language might be written off as a mistake or an aberration. But four bills introduced within months of each other that could enable cloned fetal farming suggest a pattern of premeditation and intent. It seems that the biotechnology industry is unwilling to settle for any real limits on cloning research.

Why would the legislative sponsors of the New Jersey, Texas, Delaware and Maryland bills want a license to gestate cloned fetuses through the ninth month? To appease biotech companies, surely. But toward what end? Here are some possible answers to this disturbing question:

To develop cloned organs for transplantation. SCNT boosters claim that therapeutic cloning could be used to create new organs for transplantation that match the DNA of the recipient. For example, if a patient needed a new liver, he or she could be cloned, the stem cells extracted from the embryo, and then organs or organ tissue developed for transplantation into the patient. Unlike organs from other people, the theory goes, these organs or tissues would not stimulate rejection and would not require strong immunosuppressant drugs since most of the introduced DNA would be identical to that of the patient.

How would this be accomplished? Some cloning propagandists claim they could morph organs whole-cloth from cloned ESCs. But the experiments that have so far been conducted toward this end have taken a very different approach—one that could explain the expansive license granted by New Jersey law.

There was a big news story a few years ago touting a supposed therapeutic cloning breakthrough in which researchers at Advanced Cell Technology used cloned ESCs from a cow to create a nascent cow kidney. The scientists then transplanted the nascent kidney via skin graft onto the cow that had supplied the DNA. The alleged advance: the kidney produced urine and the cow's body did not reject the transplanted tissue. Here was proof that therapeutic cloning was a viable option for providing organs to human patients needing transplants. Reuters, for example, reported that the experiment proved that "therapeutic cloning will work."[61]

That isn't exactly true. Yes, a cow embryo was cloned and developed to the blastocyst stage. But the embryo was not dissected for its stem cells, as would be done in therapeutic cloning, nor were its ESCs developed into kidney tissue. Rather, the clone blastocyst was implanted into a second cow's uterus and gestated for several weeks until the fetus developed a nascent kidney. At that point, the fetus was aborted, the nascent kidney harvested and then grafted back onto the original cow. This, as the peer-reviewed journal in which the experiment was reported acknowledged, was "*not* an example of therapeutic cloning."[62] What it demonstrated was something more malign: that cloned fetuses might be gestated for use as organ donors.

To replicate this procedure for a human patient, a cloned embryo would be made using the patient's DNA in the SCNT process and then developed to the blastocyst stage. At this point, rather than harvest the embryonic stem cells, as would be the case in purely "therapeutic" cloning, biotechnologists would implant the clone in a woman's uterus (or artificial uterus if this is ever developed). The woman would gestate the embryo to the stage where the desired organ had developed sufficiently. The fetus would then be aborted and the organ harvested for transplantation. So far, no one has advertised their wish to carry out this procedure; but if it were done in New Jersey today, it would be perfectly legal.

To study abnormal embryonic and fetal development. Researchers may wish to create human clones with predefined genetic defects to study certain conditions and their causes. For example, clones could be made from the DNA of anencephalic babies (i.e., lacking most of the brain and spinal cord) in order to allow the defective fetuses to develop, and then be aborted in order to study how this deformity arises. Indeed, this very prospect was raised at the neuroethics conference, discussed earlier, at which Henry T. Greely said that California's fourteen-day legal limit for maintaining cloned embryos could one day be extended.[63]

To learn how to engage in cloning-to-produce-children. Recall from the last chapter that a primary cause of death or poor health in cloned mammals seems to be defective gene expression during embryonic and fetal development. Learning how to overcome these problems would be a daunting task that could not be limited to experiments on early embryos in Petri dishes, analyses of simple cell lines or computer mockups. As described by David Prentice in the last chapter, it would eventually require scientists to transfer clones into the uteruses of women volunteers (or into artificial wombs) to analyze early embryonic development and ensure that the gene expression problems had been overcome. "Each of these clonal pregnancies would be terminated at various points of development, each fetus destroyed for scientific analysis."[64] Such work would not be legal under the Hatch/Feinstein approach to permitting human SCNT, since this would require destruction of the clone after fourteen days and would prohibit implantation. But it would be explicitly legal under the cloning license granted to Big Biotech in New Jersey.

For testing of new medicines. In the more distant future when cloning techniques may be more reliable—and especially if artificial wombs are ever developed—cloned fetuses deliberately manufactured to develop certain diseases might be used to test new pharmaceuticals for efficacy and safety, perhaps replacing the use of animals in medical research. This would fit snugly with the agendas of some bioethicists and members of the radical animal rights movement, who seek to reduce or eliminate all medical testing on animals. Some of these thinkers have even suggested that cognitively devastated people should be used in medical experiments to spare the primates.[65]

■ ■ ■

Is a cloning compromise possible? So we come to a bedrock question: If a total legal ban can't be enacted, would it be better to accept a phony ban that explicitly authorizes human SCNT for research purposes, or to have no law at all? That's the tactical problem that seems to be facing cloning opponents, but in actuality it's no problem at all. There's little to be gained from such legislation, and much to lose.

Why?

First, because such a law will be more likely to lead directly to reproductive cloning than actually prevent it: with the research legalized, more effort will be put into learning how to reliably clone human life.

Second, a law authorizing human SCNT will open the spigot of venture-capital investment that biotech companies and cloning researchers desperately crave. Once the serious money begins to flow, it will vastly increase the likelihood that successful human cloning techniques will be developed.

Third, it will permit the intentional blurring of the distinction between ESC research from IVF embryos and the creation of human cloned embryos via SCNT. This will eventually lead, as it has already in California, to laws permitting the government to fund human cloning research. As we have seen, California cloning lobbyists are pushing a voter initiative that would require the state to borrow $3 billion over ten years, part of which would be used to pay for human cloning research.[66]

Fourth, for reasons we have already pointed out, it is very unlikely that "therapeutic cloning" will ever become a widely available treatment for degenerative conditions. And while there may be other uses for CBR—such as creating new embryonic stem cell lines that embody specific genetic diseases, in order to study how these diseases affect embryonic development—that is not how cloning has been sold to the American public, nor is it sufficiently persuasive to overcome the ethical impediments to permitting human cloning.[67]

Fifth, since cloned mammals have defects, some scientists worry that these defects could also affect the safety of cloned stem cells used in medical therapies.[68] Whether this is true would require years of determined research.

Sixth, license to engage in human SCNT will quite likely lead to conducting immoral research on cloned human fetuses. Moreover, such experiments will be essential in learning how to engage in genetic engineering and the new eugenics of creating "designer babies"—a prospect, as we shall see, that is already being actively promoted by some bioethicists and futurists.

Finally, such a law will divert resources best focused on the more promising ASC research and other regenerative medical techniques.

The only way to accomplish the goal of developing regenerative medicine—without at the same time creating a superhighway leading to a Brave New World—is to outlaw all human SCNT. This idea is not antiscientific or Luddite. One need not be a religious conservative or pro-life on abortion to support such a law. After all, it has been adopted by many nations in the world, including "progressive" ones such as Canada, Australia, Norway, and France. Internationally, Costa Rica is leading a drive to obtain agreement on a complete ban in the United Nations.[69] Anything less would not be "compromise"—it would be surrender.

CHAPTER 5

Political Science

. . . .

"THERE ARE MANY BUSINESSES that involve science," said Tom Abate, author of *The Biotech Investor,* when I asked him what makes biotechnology different from other profitable enterprises that involve science and medicine. "The whole of Silicon Valley is based on electrical engineering, solid state physics, materials science, chemistry and the like." He then identified why biotechnology has become as controversial as any field of research in modern times. "The difference is that none of these others involve questions of life. As soon as you have a question of life you have inserted moral controversy that is absent in other fields."[1]

While much of biotechnological research is wholly uncontroversial—for instance, recruiting adult stem cells to rebuild a patient's damaged organs—I believe that he understated the case. Biotechnology involves life, but it also involves altering the natural biology of plant, animal and human life. No wonder biotech is such a deeply problematic enterprise.

The vast majority of current biotechnological innovations involve bioengineering *plant or animal* life for human benefit, undertakings I find relatively untroubling, but about which many environmentalists are up in arms. I am most concerned about endeavors that directly apply to *human* life. Indeed, the controversies with which we are grappling here open the possibility that scientists will someday be able to alter human life at the molecular level, transform some human lives into manufactured products, and given sufficient time, perhaps even engineer the genome so radically that some people will no longer be human, as we use the term today.*

At the same time, we must acknowledge that biotechnology offers tremendous potential to eradicate disease, extend lives and alleviate agony, and—let me make myself very clear here—most of the people in the field, including those with whom I profoundly disagree, truly hope that their work will bring succor to a suffering world.

Perhaps this desire explains why so many scientists, biotechnologists and bioethicists often lash out so emotionally and bitterly at critics of unfettered research. But maybe there are other reasons as well. The vehemence with which many biotech boosters attack critics seems far too emotional to be merely expressions of frustration at having the moral value of their work questioned. Rather, the level of vituperation and demagoguery aimed at "the opposition" seems more a reflection of a passionate ideological commitment to research for its own sake, causing scientists to resent bitterly any critics who reject their "science *über alles*" views.

University of Pennsylvania bioethicist Art Caplan, for example, castigated opponents of human cloning as "a bizarre alliance of antiabortion religious zealots ... technophobic neoconservatives, [and] scientifically befuddled antibiotech progressives," whom he unfairly accused of "pushing hard to ensure that ... more moral concern" is given "to cloned embryos in dishes than ... to kids who can't walk and grandmothers who can't hold a fork or breathe."[2]

Caplan's colleague Glenn McGee, who is editor-in-chief of the *American Journal of Bioethics* and a rising star in the bioethics movement, recently matched Caplan's invective when he compared Leon Kass, the chairman of the President's Council on Bioethics, to a fictional movie assassin.

McGee was apoplectic because Kass has powerfully inserted the issue of human dignity into the debates over human cloning, genetic engineering and the manufacture of human/animal hybrids (chimeras). So he lashed out at Kass for "putting a stop to embryonic stem cell

*For readers who want a broader perspective on the breadth and scope of these areas of biotechnological research, I suggest reading Michael Fumento's splendid *BioEvolution: How Biotechnology Is Changing Our World* (San Francisco: Encounter Books 2003).

research, and if possible, putting a stop to a number of other scientific and clinical projects objectionable to the far-right wing of the Republican Party, and in particular, Southern Baptists."[3]

This wasn't rational debate; it was diatribe. As McGee well knows, Kass is Jewish, not Southern Baptist. Moreover, neither Kass nor President Bush has ever advocated outlawing embryonic stem cell research. (Both do oppose human cloning, which is not the same thing as embryonic stem cell research, although many cloning advocates strive mightily to blur the distinction.) Indeed, Kass has never publicly endorsed the president's August 2001 decision to permit limited federal funding on existing embryonic stem cell lines.

Most of us who oppose human cloning are enthusiastic believers in scientific progress. But we also recognize the dangers inherent in a philosophy of "anything goes." And because biotechnology has developed the power to tamper with human nature at the molecular level, we believe that scientists have a profound responsibility to exercise prudent self-restraint, and furthermore that society owes future generations the setting-up of reasonable regulations to protect their genetic integrity. Kass expressed these points quite well in *Life, Liberty and the Defense of Dignity:*

> Because it is essentially *instrumental,* technology is itself morally neutral, usable for both good and ill. There are, of course, dangers of abuse and misuse of technology, but these appear to be problems not of technology but of its use human users, to be addressed by morality in general. And, besides abuse and misuse, there is the genuine problem of technology itself: the unintended and undesired consequences arising from its *proper* use. Thus, the problems of technology can be dealt with, on one side, by technology assessment and careful regulation (to handle side effects and misuse), and, on the other side, by good will, compassion, and the love of humanity (to prevent abuse). This combination will enable us to solve the problems technology creates without sacrificing its delightful fruits.[4]

What part of this perspective could provoke such outbursts by some of those who disagree?

Perhaps it is the call to love humanity for its own sake. McGee's accusation concentrates on Kass's belief that it would be repugnant to manufacture human clones in laboratories or breach natural species boundaries through the fabrication of human/animal chimeras, and on

his repeated advocacy over many years of the inherent dignity and special meaning of natural human life—in front of which, Kass has often
stated, "we should stand in awe."[5]

Declarations of human exceptionalism are like a stake through the
heart to many bioethicists, futurists, and members of the scientific and
medical intelligentsia. Kass is widely castigated by many among the
bioethical and biotech elite precisely because he is viewed—correctly,
in my view—as our premier apologist for human uniqueness. This makes
him Enemy Number One. For if law and social morality adhere to the
principle that human life has special value simply because it is human,
many Brave New World agendas will be thwarted. It is precisely because
Kass has so eloquently warned against the consequences to humanity
of engaging in these technologies that McGee accuses him of leading "a
new anti-science elite" that seeks to impose "a neoconservative natural
law theory" upon science when surely every ignoramus knows that
human "nature doesn't really exist," or at least "can't be operationalized
in science or policy."[6]

This us-versus-them mentality, by politicizing science and the research
enterprise, has introduced an unhealthy politico-scientific correctness into
the entire debate. Matters have gotten so bad, in fact, that some researchers
tailor their published data to influence the political impact their findings
might have on the controversies raging around biotechnology.

Rick Weiss, science editor of the *Washington Post,* who is certainly
no enemy of embryonic stem cell (ESC) research or human cloning, presented one example of this phenomenon. During the debate over federal funding of ESC research early in the Bush administration, researchers
planned to report that mouse embryonic stem cells harbor genetic abnormalities. But such findings, at a moment when the controversy was reaching a crucial point, might have strengthened the case of those arguing
against federal funding. So, at the last minute before publication, the
researchers revised their report: where the original version had "called
for research" to see if the "genetic instability" of ESCs might "limit their
use in clinical application" (that is, as an efficacious treatment), the
rewritten paper simply extolled the potential therapeutic value of ESCs
and left it at that. Editing out an important warning that more research
was needed was certainly not science. It was politics.[7]

The same mentality was on display in an editorial published in the
The Lancet with the unconsciously ironic title, "Facts versus Ideology

in the Cloning Debate," protesting the media's reporting that South
Korean researchers had successfully created thirty cloned human
embryos.[8] Even though Woo Suk Hwang and his colleagues had acknowledged in their report, published in the journal *Science,* that "25 percent
of the embryos" they had created through SCNT had reached "the blastocyst stage," *The Lancet's* editorial writer, twisting semantics, declared
otherwise.[9] Rather than being cloned embryos developed to the blastocyst stage, the entities in question were merely a "line of undifferentiated, pluripotent cells whose 'parents' were an unfertilized oocyte . . .
and a cumulus [ovarian] cell." Hence, the editorial pontificated, Hwang's
successful acts of human cloning hadn't really created human life: "No
one made a new human baby, let alone destroyed it. No one made an
embryo as most people understand the term, derived from the fusion of
an oocyte and a sperm."

Actually, this is nonsense. A blastocyst is an embryo and embryonic stem cells by definition come from embryos; this is basic biology.
Since when do prestigious medical journals reject correct scientific definitions in favor of those based on how "most people understand the
term"? What's more, Hwang admitted in several media interviews that
the SCNT cloning technique "cannot be separated from reproductive
cloning of people," which would start with the creation of an embryo.[10]

"Left where they were," *The Lancet* continued, "neither the cumulus cells nor the oocytes could have become separate human beings."
Obviously not. This is like stating that "neither sperm and eggs, if left
alone, could have become separate human beings." It's utterly elementary that mere cells will never gestate into babies—a biological feat that
requires an embryo. Indeed, in a sudden semantic U-turn that belied all
his previous assertions, the editorialist admitted that "such an artificially
created blastocyst, if implanted back into a womb, could perhaps have
developed further, eventually to a new person—that is, reproductive
cloning."[11] In other words, the nonembryos turned out to be embryos
after all. So it wasn't the popular media that had engaged in "ideological posturing" when it reported Hwang's clones to be embryos. The ideological spinners were *The Lancet's* editors.

Since objectivity is one of the bulwarks of the scientific method,
it's surprising to see a leading journal of medical science compromise
this principle. Moreover, *The Lancet* is not alone in doing so. A 2003
editorial written by the *New England Journal of Medicine* editor-in-chief

Jeffrey M. Drazen, for example, asserted that since it is "unreasonable to prohibit research" on human SCNT cloning, the editors of the *NEJM* vowed to do their part to influence the political debate "by seeking out highly meritorious manuscripts" that extol the virtues of ESC research and human SCNT.[12] So the *NEJM* wasn't merely going to editorialize about the issue, which would be perfectly appropriate, assuming that they based their opinions on scientific facts. Rather, the selection of articles for publication would be based, at least in part, upon whether they promoted the *Journal's* stated goal of "deterring political opposition to research."[13]

This politicization of editorial decision-making raises several unsettling questions: What if the *Journal* received a credible paper describing a major ASC research breakthrough that called into question the need to harvest tissues from cloned or natural embryos? Is it unreasonable to wonder whether the editors, knowing that doing so might harm their avowed political agenda of legalizing human cloning-for-biomedical-research (CBR), would publish it?

Or, what if the *Journal* received a manuscript reporting that an attempt to use cloned ESC therapy in mice to treat diabetes had failed? And what if this came at such a delicate moment in the political debates that a candid report about the failed experiments might increase the chances that "unreasonable" legislation would pass, prohibiting CBR? Or what if a submission for publication indicated that ESCs' known propensity to cause tumors when injected into animals might be an insolvable problem? What then? Publishing such an article would unquestionably compromise the editors' plans.

Drazen's grossly inaccurate description of the science of human cloning magnified these worries:

> There are two distinct uses of embryonic stem cells. The first, for which there is no support among members of the scientific and medical communities, is the use of stem cells to create a genetically identical person. There is a de facto worldwide ban on such activities, and this ban is appropriate. The second use is to develop genetically compatible materials for the replacement of diseased tissues in patients with devastating medical conditions, such as diabetes or Parkinson's disease. This is important work that must and will move forward.[14]

It is hard to believe that the editor-in-chief of one of the world's most prestigious medical journals thinks that an "embryonic stem cell" could create a "genetically identical person," a reference to the birth of a cloned baby. This is scientifically illiterate. Stem cells are merely cells; implanting them could no more lead to a pregnancy than placing a blood cell or a skin cell into a woman's womb.

Not only that, but SCNT does not produce stem cells *per se*; if successful, it produces *cloned human embryos,* from which stem cells could be extracted. But these same embryos could also be used to create a "genetically identical person" if implanted into a woman's womb and gestated until birth. While a stem cell is just a cell, an embryo is a distinct, individual human life, albeit in a nascent stage of development. In the name of scientific accuracy and integrity in advocacy, Drazen should at least have made these fundamental biological distinctions clear.

Judgments about human cloning and other biotech agendas are not just scientific issues; they belong also to the realms of philosophy, ethics, religion, values and ideology. But moral reasoning on these subjects must begin with a clear understanding of the science involved. Unfortunately, in order to win the human cloning debate, many scientists seem willing to compromise the unwavering objectivity that ideally should separate science from politics. As Roger Pielke Jr., director of the Center for Science and Technology Policy Research at the University of Colorado, warned in *Nature,* "Many scientists [now] willingly adopt tactics of demagoguery and character assassination as well as, or even instead of, reasoned argument" in promoting their views. This politicization of science has led some scientists, not to mention "lawyers and those with commercial interests [to] manipulate 'facts' to support their advocacy, undermining the scientific community's ability to advise policy makers." As a consequence, science "is becoming yet another playing field for power politics, complete with the trappings of political spin and a win-at-all-costs attitude."[15]

This pathological trend has gone so far that, in order to influence the making of public policy, a few scientists have even resorted to misrepresenting their research. The most notorious such case occurred in Australia, where Alan Trounson, a leading stem cell researcher, admitted to releasing a misleading video so as to "win over politicians" during that country's parliamentary debate over ESC research. The video depicted a disabled rat regaining the ability to walk (or so Trounson claimed) after being injected with ESCs. In actuality, the experiment used cadaveric fetal tissue from five-to-nine-week-old aborted human fetuses, an altogether different approach that was irrelevant to the ESC debate.[16] Worse, it later turned out that the rat had been "helped," but not "cured."[17]

More common than outright lying is the cruel practice of giving false hope to people by allowing them to think that cures are just over the horizon when they are probably more than a decade away—if they are coming at all. A notorious recent example of this occurred after Ronald Reagan's death from complications connected with Alzheimer's disease. Nancy Reagan, apparently believing that ESCs could help treat Alzheimer's, made headlines when she urged President Bush to loosen the restrictions he had placed on federal funding.

But it turns out that ESCs will almost certainly never be an effective treatment for Alzheimer's because, as two respected experts told a Senate subcommittee, it is a "whole brain disease" rather than a cellular disorder such as Parkinson's.[18] In the words of science correspondent Rick Weiss, "stem cell experts confess ... that of all the diseases that may be someday cured" by stem cell treatments, "Alzheimer's is the least likely to benefit."[19] (This verdict is very likely true of ASC therapies as well.)

But there are risks in allowing public policy debates to be overly influenced by the positions taken on them by famous people. Scientists and ESC propagandists have allowed celebrities (like the Reagan family) to believe that ESCs could cure Alzheimer's—"a distortion that is not being aggressively corrected by scientists."[20] Why? Simply because the false story line helps generate public support for the biotech agenda. "People need a fairy tale," said Ronald D. G. McKay, a stem cell researcher at the National Institute of Neurological Disorders and Stroke. "Maybe that's unfair, but they need a story line that's relatively simple to understand."[21]

What could explain such blatant, unashamed politicization of science? Roger Pielke thinks that some people simply don't understand the difference between "scientific results" and "the policy significance of those results." This may be true, but is Pielke giving the matter short shrift?

The research programs on the frontier of biological science have become potent *symbols* of the ongoing "culture wars" that divide our society. Underlying the rhetoric and the political maneuvering are the deep tensions between materialism and religion, liberal and conservative, abortion rights and pro-life, relativism and traditional values. On one side of the political battlefield, there are those who are particularly intense because they "see science as having a mission that goes beyond the mere investigation of nature or the discovery of physical laws. That mission is to free mankind from superstition in all its forms . . . ."[22]

"Scientism" Isn't Science

Science, properly pursued, should be apolitical. Ideally, as dispassionate users of the scientific method (including observability, testability, repeatability, peer review), scientists are objective in their pursuit of knowledge. While scientists are certainly entitled to express opinions about what they perceive to be sound public policy, their job in their scientific work is to find the facts and let the political chips fall where they may.

This time-honored stance of scientists has served us well and the practical results are all around us. With the help of applied science, we have risen some distance above the unforgiving tooth-and-claw of brute nature. Simply stated, science is integral to achieving progress and improving the human condition.

But there is, of course, a potential downside to science: it confers on us dangerous godlike powers to destroy as well as create. For this reason alone, wisdom prescribes that we establish checks and balances to ensure that the scope of research remains consistent with the values of liberty, justice and enlightenment. As Francis Fukuyama put it so well in *Our Posthuman Future*, "True freedom means the freedom of political communities to protect the values they hold most dear, and it is that freedom that we need to exercise with regard to the biotechnology revolution today."[23]

Unfortunately, many devotees of biotech increasingly resist any meaningful social restraint; they perceive science not just as a means for obtaining knowledge to promote the common good but as an end in itself. As a result, the scientific establishment is becoming ever more insulated from the rest of society. In a disturbing admission that reveals this professional deafness, Bruce Alberts, president of the National Academy of Sciences, told a reporter for the *National Journal,* "We [scientists] care a lot about how other scientists think about us, and we don't care a lot about others who are not scientists."[24]

For some, this wholehearted belief in science as an end rather than a means has become something of a pseudoreligion, an ideology known as "scientism." As one online science encyclopedia defines it, scientism "claims that science alone can render truth about the world and reality.... Scientism sees it as necessary to do away with most, if not all, metaphysical, philosophical, and religious claims, as the truths they proclaim cannot be apprehended by the scientific method. In essence, scientism sees science as the absolute and only justifiable access to truth."[25]

What does this mean in practical terms? In the words of particle physicist Stephen M. Barr, "There is more to materialism than ... cold ontological negation. For many, scientific materialism is not a bloodless philosophy but a passionately held ideology. Indeed, it is the ideology of a great part of the scientific world."[26] If religious fundamentalists believe that all the answers come from the Bible or the Qur'an, then science's ideologues place all their trust in the book-in-progress of science. As Stuart Newman, a professor of cell biology and anatomy at New York Medical College in Valhalla, described it to me:

> Scientism takes the methods of experimental and theoretical science and says that's what you need to understand everything. You look at the successes of physics, chemistry and certain kinds of molecular biology, and then say this should be the paradigm for our understanding of *everything*. And we shouldn't impede *any* kind of science activity because science has shown itself to be so beneficial. That's moral scientism.[27]

Along these lines, one British true believer succinctly described her devotion to scientism, writing that "the notion of a universal moral law is obsolete: We take our moral laws today from science and the state."[28] Rahul K. Dhanda is even more blunt:

> Whether science's pursuit of knowledge ought to eclipse the interests of society (scientists would argue that their work will ultimately aid society) remains less an issue than whether scientists know better than society. Authorities in biology, for instance, often take the stance that their knowledge affords them the better vantage point in social discussions, thus the layperson opposed to progress ought not to weigh in on the debate. Put another way, science knows what is good for society like a parent knows what is good for the child.[29]

And Michael Shermer, writing in *Scientific American,* warns us that scientism has become a quasi religion in which naturalistic answers provide spiritual sustenance that believers do not otherwise find in religion or philosophy, and for which leading scientists are the cultural avatars and "premier mythmakers of our time." They are turned into shaman-like figures: "[T]his being the Age of Science, it is scientism's shamans who command our veneration."[30] With this sort of emotional commitment in the air, it's not surprising if many in the scientific establishment now proclaim an almost unlimited right to what they call a "freedom of science," supposedly "derived from freedom of thought and expression."[31]

According to the devotees, scientific endeavors may not be restrained except for extremely compelling reasons—reasons to be determined, naturally, only by scientists themselves. Scientists alone are entitled to judge the morality of their experiments—even if tax money is being used, even if the experiments collide with views held dear by the general public. A science professor's angry correspondence claiming a "privileged position" for professionals in determining what should and should not be done, published in *Nature* in 2003, epitomizes this outlook:

> The reactions of non-specialist observers to complex ethical problems raised by cutting-edge science such as embryonic stem cell research are no more justified or useful than their opinions about the technical difficulties yet to be overcome.... Scientists must take the lead in ensuring that the progress of science is both ethical and as free from political intervention as possible, if for no other reason than only they can do so.[32]

In other words, laypeople who are not part of the science establishment or who do not share its predominant moral ethos should mind their own business, continue to contribute their money, and shut up.

What's more, the moral views of our scientist betters are often incompatible with the general sensibilities of society. "Most scientists adopt utilitarian perspectives on ethical and political questions, and they use their values to estimate costs and benefits," the National Academy of Science's Bruce Alberts admitted to the *National Journal.*[33] He acknowledged that the public's scruples sometimes prevent research that scientists' predominantly utilitarian outlook would permit them to pursue with few qualms. For example, in the 1970s a public outcry halted the practice of using condemned prisoners in experiments, whereas "If it was purely up to the scientists, they might accept the idea of doing experiments on death row [because] the person will be dead in six months anyway."

The intellectual and ethical values of the science establishment are so uniform, according to Alberts, that scientists worldwide constitute "a special clan, a family who have special allegiance to each other." This clannishness may even extend to punishing other scientists who are perceived to have violated the "clan's" values and mores by dissenting from the party line.[34] Dissenters often find it difficult to be published in mainstream journals, obtain tenure at universities, receive research grants, or be invited to share their views in symposia—all of which are required to advance a science or academic career.

David Prentice knows how it feels to be ostracized and condemned for voicing heterodox views. As cofounder of Do No Harm: The Coalition of Americans for Research Ethics, he is one of the few life scientists willing to advocate outlawing human cloning-for-biomedical-research publicly.

Prentice's approach in presenting his views is entirely empirical. He reviews the plethora of peer-reviewed journal articles published on cloning and stem cell research and uses the information he gleans to mount compelling public presentations. One of the most persuasive of these clearly demonstrates how, in regenerative medicine experiments, researchers have so far achieved superior results with ASCs and other nonembryonic sources over those derived from embryos, whether natural or cloned. One is free to disagree with Prentice, but there's no doubting that his work displays investigative rigor and intellectual integrity, because of which he has served as a science advisor to Senator Sam Brownback (R-KS) and Representative Dave Weldon (R-FL). He has been invited to present his views throughout the world to governmental committees and in public forums, and even as an official advisor to the Costa Rican diplomatic delegation to the United Nations, where he aided Costa Rica's efforts to gain an international ban on all human cloning.

But Prentice's endorsement of a formal prohibition of human cloning is heresy to the science and bioethics establishments. Unable to rebut his assessment of the science, his critics have instead frequently resorted to personal attacks, innuendo and smears against his academic and scientific credentials. An article in the bioethics journal *Hastings Center Report* accused him of using "tobacco industry" tactics and of "devaluing scientific knowledge." Why? Because Prentice says that ASC research has advanced further than ESC research and promises as much as therapeutic cloning, but without the moral onus.[35] Similarly, after Prentice testified in favor of legislation that would have banned human cloning in Nebraska, an advocacy group accused him of not actually being a stem cell scientist.[36]

In what appears to have been an attempt to overburden him with work so as to reduce his effectiveness as a critic of biotech, Prentice's teaching load at Indiana State University was substantially increased (after he returned from a sabbatical spent in anti-cloning advocacy), from a normal three classes with 30 students in spring semester 2001 to six classes with 193 students in fall semester 2002. Three of the latter were large freshman classes that a professor of Prentice's stature and years of service does not usually teach. Indeed, before he became a controversial figure within the scientific community, Prentice's classes were small and restricted to biology majors or graduate students.[37] He believes that being forced to teach two or three "punishment" classes each semester was revenge for defying the party line on cloning and stem cells.*

Other dissidents within the biotechnology community have been forced into the closet of silence for fear of having their careers ruined by the science thought police. As the *National Journal* reported, "Several researchers said they dared not comment about this debate because they feared informal retribution from colleagues who can withhold the support necessary to get jobs, grants, or publishing opportunities."[38]

This repressive atmosphere was illustrated in a 2004 article in the *San Francisco Chronicle* by science journalist Mark Dowie, aptly titled

---

*As I was completing work on this book, I discovered that Prentice had decided to leave Indiana State University to accept a position in a think tank in Washington, D.C. I asked him, "Why allow yourself to be forced out, considering you have tenure?"

"Having tenure means they can't fire you, but they can 'encourage' you very strongly to leave," he told me in explaining his decision. "It was pretty clear that I was no longer wanted by my department and I grew very tired of the fighting, the continual tension, and the petty harassment, difficulties that had not existed before I began my public work. So I decided to go where my efforts will be appreciated. Frankly, it's a relief. I haven't been this happy in years." Prentice will be the senior fellow for life sciences at the Family Research Council. (David A. Prentice, interview with author, June 9, 2004.)

"Biotech Critics at Risk." Dowie decried "a disturbing trend in modern science" in which scientific and academic freedom are eroding in the face of economic and peer pressures. Researchers who have "challenged the catechism" of current understandings in biotechnology have been attacked by their colleagues, lost their jobs, had grant funding stripped and their research "stifled," and have been subjected to "an orchestrated international campaign of discreditation."[39] This shameful social phenomenon might best be described as "secular excommunication."

The clique mentality has apparently gotten so bad that those outside the science "clan" may not even bother to apply for NIH and other grants. Dr. Michel Levesque, the neurologist we met in the first chapter who may have successfully treated Parkinson's disease with the patient's own brain stem cells, did not apply for an NIH grant because, he told the *National Journal,* "it is like a fraternity more than anything else. . . . It is a buddy system." Displaying the mindset, James Battey, head of the NIH Institute of Deafness and Other Communication Disorders, haughtily dismissed Levesque's concerns, likening him to a member of the Raelian cult.[40]

Scientism not only undermines the objectivity and intellectual rigor of science, but also inculcates a dangerous utopianism. The eugenics movement of the early to mid-twentieth century is a classic case in point. Eugenics—literally, *good in birth*—was the brainchild of Charles Darwin's cousin Francis Galton, who sought to apply Darwin's theory of evolution and Mendel's discoveries in genetics to improve human society by enhancing the gene pool. The result was a slow crescendo to horror: tens of thousand of Americans involuntarily sterilized, with similar actions taken in Canada, Australia and the United Kingdom. In Germany, of course, eugenics fueled the fires of the Holocaust.

The beginning was seemingly benign. Galton hoped to apply "positive eugenics"—that is, the encouragement of those he deemed to have the best genetic stock to marry and procreate bountifully. Who could be hurt by that? But utopian ideas, once unleashed and made respectable with the official approval of science, are not easily contained. Eugenics ideology was an infectious disease that spread quickly to the United States, its triumph hastened by the transparent monomania of scientism. By the early 1900s, it had been enthusiastically embraced by members of the upper crust. Its boosters received generous financing to promote their ideas throughout the country, most particularly from the Carnegie Institution. "The men and women of eugenics wielded the science," Edwin Black writes in his superb history of the American eugenics movement, *War Against the Weak.* "They were supported by the best universities in America, endorsed by the brightest thinkers, financed by the richest capitalists."[41]

Instead of focusing primarily on promoting "good" marriages, early eugenicists like Charles Benedict Davenport vigorously promoted "negative eugenics," aimed at "redirecting human evolution" by legally preventing—by means of forced sterilizations—the "unfit" from reproducing. Toward this end, Black points out, "esteemed professors, elite universities, industrialists, and government officials," relying on biological rationales,[42] unleashed a sterilization pogrom to cleanse the gene pool of the "feebleminded, the pauper class, the inebriate, criminals of all descriptions, including petty criminals ... epileptics, the insane, the constitutionally weak ... those predisposed to specific diseases, the deformed, and those with defective sense organs, that is, the deaf, blind, and mute."[43]

By 1910, "eugenics was one of the most frequently referenced topics in the *Reader's Guide to Periodic Literature.*"[44] In its boom years in the 1920s, eugenics became a serious, scientifically supported and influential social and political movement. Courses in eugenics were taught in more than 350 American universities and colleges, leading to widespread popular acceptance of its pernicious tenets.[45] It was endorsed by more than 90 percent of high school biology textbooks.[46] Eugenicist societies formed to promulgate and discuss the theory and academic eugenics journals sprouted. What is more, philanthropic foundations embraced the movement, financing research and policy initiatives. Many of the most notable political, cultural and arts figures of the era believed in it—including Theodore Roosevelt, Winston Churchill, George Bernard Shaw, Clarence Darrow, Helen Keller and Margaret Sanger—and their opinions could only reinforce the movement's popularity.

Eugenics reached hurricane strength after the U.S. Supreme Court sanctioned forced sterilization as a public good in 1927.[47] About 6,000 eugenic sterilizations took place in the United States between 1907 and 1927. By 1940, the number had climbed to nearly 36,000. By the time the practice ended in this country in the early 1970s, nearly 70,000 of our fellow Americans had undergone the operation, all under the mandate of law.[48]

It could have been even worse. "Murder was always an option," Black writes.[49] It didn't take long for some eugenicists to see that euthanasia of the unfit, in addition to sterilization, would be a splendid means of preventing the spread of genetic pollution.[50] Thankfully, this final solution was not undertaken officially in the United States, as this country, to its credit, shrank from carrying eugenics theory to the logical conclusion pursued by the Nazis.

Utterly discredited by the carnage of the Holocaust, eugenics had seemingly died during World War II. But it was only hibernating. It has since reawakened, Black warns, in the guise of a utopian "newgenics" advocated by "self-ordained experts" in bioethics and bioscience who urge that we "harness the nature-changing power of genetics and the energy of entrepreneurial enterprise to once again chase in vain after the mirage of human perfection."[51] The growing social movement known as "transhumansism" advocates the technological refashioning of humanity and the creation of a "posthuman" species. At the heart of its program is the legalization of human cloning. Steeped in scientism, transhumanism is explicitly eugenic in outlook. Indeed, as Black says, "Prominent voices in the genetic technology field believe that mankind is destined for a genetic divide that will yield a superior race or species to exercise dominion over an inferior subset of humanity."[52]

We don't have to go down this path, of course. Most people are undoubtedly repelled by the very notion of transhumanism. But if we are to contain the dangerous tendencies of scientism, we can't leave the moral judgments just to scientists.

The Scientific-Industrial Complex

Scientism's ideological power is strengthened by an extravagantly financed political campaign on the part of the biotech industry to block or obliterate legal impediments to human cloning and other Brave New World technologies.

"Most scientists believe that their pursuit of objective truth is the sole consideration accepted by their profession," Sheldon Krimsky writes in *GeneWatch,* published by the Council for Responsible Genetics. "Whatever other interests they may have in the subject matter of their research [e.g., financial, ideological] are believed to be eclipsed by, and subservient to, the unfettered search for certifiable knowledge." This may have been true in the past, but no more. Sadly, Krimsky notes, "Academic science ... became intensely commercialized during the last quarter of the twentieth century, a result of complex events including new laws, court decisions, executive orders, and growing incentives among research universities partnering with the private sector."[53]

Many biotech players, confident that the emerging life science industry will be the next Silicon Valley, have huge dollar signs swimming in their heads. There's nothing wrong with wanting to be rich, of course. But the lure of great wealth provides a powerful incentive to hold self-justificatory beliefs and embark on morally questionable endeavors. Indeed, the biotechnology industry has coalesced into one of the nation's most powerful special interests—Big Biotech—which spends tens of millions of dollars each year in contributions to public relations and political campaigns. It works tirelessly to influence public opinion and convince government officials and regulators to accept the official biotech view that—with the exception of (for now) bringing a cloned baby to birth—researchers should be allowed virtually a free hand.

Patent or Perish

Big Biotech has projected itself into the nation's universities, causing marketplace values to corrode intellectual standards. Being published, and being subjected to vigorous peer reviews in professional journals, does remain an important part of contemporary academic science. But these days, Ph.D.'s may be more likely to rush to the patent office in the hope of selling their findings to a biotech or pharmaceutical company than to share them with colleagues in a learned journal.

This novel development represents a radical shift in the ethical assumptions that have long governed academic research. Imagine this scenario: Albert Einstein, while at Princeton, makes a world-shaking theoretical breakthrough. But he wants more than the joys of discovery, plaudits from his peers and the admiration of a grateful public. Neither is he satisfied with the freedom that the university gives him to think rather than teach, nor the tenured job security and generous salary he receives. So instead of publishing in an academic journal and holding a symposium to share his theory with the world, he patents spinoffs and forms Relativity Inc. in partnership with an investment banking firm. Within a few years, the company raises hundreds of millions in its IPO and is purchased by an international conglomerate. The professor retires to a life of leisure and philanthropy, listed by *Forbes* as one of the world's richest men.

Until recently, that would have been an almost unthinkable scenario. But in today's world, universities are fast becoming adjuncts to industry, and intensely commercial incentives have been brought into milieus that traditionally have been enclaves of vocational intellectuals. Journalist Neil Munro explains the situation, with regard specifically to biotechnology, in the *National Journal:*

> In times past, the public image of a scientist was that of a lone genius, working tirelessly in the laboratory to unlock nature's secrets for the common good. But a vastly larger enterprise subsumes the modern scientist. Today, each scientist's prospects depend on the work of other scientists in related fields, and on the vast industrial enterprises that have the capital, the large testing laboratories, and the worldwide production facilities needed to get medical discoveries into the marketplace.[54]

This current state of affairs may justifiably give us reason to question the objectivity of scientists when they are in a position to pontificate. Thus, when the National Academy of Sciences established a panel of "experts" to advise government policy-makers on how to regulate human cloning, many of these were actually part of the emerging "scientific-industrial complex," whose roles as "company directors, patent holders, research managers, association executives, and applicants for government research grants" belied their status as impartial scientists.[55]

The trend toward academic-business partnerships in the life sciences is accelerating. In an era of tight government budgets, academia is becoming increasingly dependent on the private sector for financing, thereby increasing the influence of corporations upon the universities' endeavors. There's no doubt that John Gearhart and James Thomson, co-discoverers of human embryonic stem cells, truly believe that their research could lead to regenerative medical treatments; but maybe it's not irrelevant in judging their vocal support for ESC research to note that they—and their affiliated institutions, Johns Hopkins University and the Wisconsin Alumni Research Foundation, respectively—will receive substantial financial benefit from their licensing agreements with the biotech giant Geron Corporation if their work leads to profitable products.[56]

As one concerned academic research scientist warned in the *Washington Post,* the growing symbiosis between universities and corporations represents "an enormous structural change in academia" in which the role of the life sciences is being transformed from "obtaining knowl-

edge to be shared" with the wider community to developing "propri-
etary knowledge that can be owned and held confidentially."[57] Adding
heft to this concern: since 1980, according to the Association of University Technology Managers, American universities had ties with "more
than 3,000 companies. From 1999–2000 alone universities' revenues
from spinoff companies, licensing deals, and outside investment shot
from 862 million to $1.26 billion."[58] Former Columbia University bio-
chemist Erwin Chargaff, one of the founding fathers of genetic research
who "provided one of the key steps in developing a structural model of
DNA," elliptically criticized this emerging corporate hegemony over biomedical research when he quipped, "If Newton were alive today, he'd
have taken out a patent on gravity."[59]

Big Biotech and Big Money

In the last twenty-five years, private investors have poured about $100
billion into the biotech sector worldwide.[60] In 2001 alone, the industry
raised $10 billion in venture capital.[61] Not surprisingly, the Biotechnol-
ogy Industry Organization (BIO) has an annual budget of tens of millions of dollars to lobby federal and state legislators, prepare position
papers, interact with government regulators, influence the media, and
in myriad other ways coordinate public advocacy on behalf of the thousands of companies it represents.

Thousands of individual biotech companies, much like the oil
wildcatters of yesteryear, strive energetically to develop products, establish relationships with scientists and researchers in major universities,
and use these connections and sophisticated PR campaigns to attract
investors. An article in the Wall Street Journal noted, "Biotechnology
companies are essentially research and fund-raising machines devoted
to selling their scientific and medical 'story' to investors and spending
the resulting cash on laboratory studies and clinical testing."[62] Most will
probably dig dry wells and go bust. But some will hit the jackpot by
striking the modern equivalent of "black gold gushers." The rewards
could be bonanzas worth billions.

Rahul K. Dhanda, a bioethicist employed by Interleukin Genetics
Inc., a large biotech company, acknowledges this point in his candid
book, Guiding Icarus: Merging Bioethics with Corporate Interests. Dhanda
believes that the science, perhaps so powerful that it "may result in the
creation of new species," is as much "in the service of capitalism" as it

is an enterprise altruistically concerned with seeking a better world.[63] Given the enormous stakes, companies are about as eager to have their research options limited as the wildcatters of yesteryear would have been for Congress to require them to file environmental-impact studies before drilling.

Scientists, biotech executives, paid lobbyists and industry-retained PR wizards ceaselessly promote the science and morality of biotech and the prospects for developing new "therapeutic cloning" medical cures; and they strive to marginalize their opponents as environmental or religious fanatics, obstacles to the inevitable, beneficent advance of science.

Companies often release dramatic public announcements touting purported research breakthroughs without first submitting their findings to the traditional peer-review process. As Robert A. Weinberg observed in the *Atlantic Monthly*:

> For many biotech companies the peer-review process conducted by scientific journals is simply an inconvenient, time-wasting impediment. So some of the companies routinely bypass peer review and go straight to the mainstream press. Science journals, always eager for scoops, don't necessarily feel compelled to consult experts about the credibility of industry press releases. And when experts are consulted about the contents of a press release, they are often hampered by spotty descriptions of the claimed breakthrough and thus limited to mumbling platitudes.[64]

Companies wouldn't engage in such hucksterism if it didn't usually work. Unfortunately, the mainstream press generally accepts these announcements without asking serious questions. Many ideologically motivated science reporters are likely to give automatic support to alleged "breakthroughs," the more controversial the better, rushing to dismiss objections to them as religiously motivated. Thus Rick Weiss, the science editor for the *Washington Post* and a self-proclaimed atheist, notoriously implied that opponents of human cloning were "religious fundamentalists" akin to the Afghan Taliban.[65]

The biotech company Advanced Cell Technology (ACT)—whose CEO Michael West is the industry's most public booster of human cloning—has been very successful at garnering free, high-visibility media promotion. ACT has repeatedly been the subject of dramatic, behind-the-scenes stories in the popular press, which reverberated worldwide in television and wire service news reports.[66]

ACT scored a public-relations coup when the December 3, 2001 issue of *U.S. News and World Report* published a nine-page puff piece extolling the alleged creation by the company of the first human cloned embryo. ACT had secretly permitted reporter Joannie Fischer to look over the shoulders of its researchers as they attempted to initiate human life through SCNT. Perhaps Fischer became too familiar with her subjects, for in her piece she wrote as an enthusiastic lobbyist rather than an objective journalist, breathlessly asserting that ACT's achievement "will be hailed as the hugest medical breakthrough of the last half century."[67]

From her embedded post at ACT, Fischer all but canonized the men who, she claimed, had created the first human cloned embryo: Jose Cibelli—the instigator—who first foresaw therapeutic cloning as "the future of medicine;" and CEO Michael West—the visionary—who seized on the concept of allowing patients to have access "to their very own cells" as regenerative medical treatments. (This, as we've seen, is scientifically inaccurate.) Finally, Robert Lanza—the activist—described by Fischer as "a living embodiment of the character played by Matt Damon in the movie *Good Will Hunting*."

Throughout her piece, Fischer repeatedly and unquestioningly parrots Big Biotech's party line. Cloned embryos "are just a few reproductive cells," West is allowed to say, that "are not much different than eggs or sperm." And Lanza: "Research advances are making all cells embryonic. But if you consider these cells human life, then 100 souls are lost every time I sneeze."

Ignoring the impressive advances that even then had been made in ASC research, Fischer quotes Cibelli on therapeutic cloning: "It's just that there is nothing else in all of medical research that is anywhere near this promising." Then she trots out Ronald Green—"chairman of ACT's bioethics committee"—who bestows the official sanction of the professional bioethicist upon the cloning project: "To commit ourselves morally to protecting every cell in the body would be insane," he says, as if anyone ever advocated such a course.[68] The reader is not told that Green has long been an enthusiastic supporter of all manner of research involving the destruction of embryos.[69]

And here's the money quote of the story—literally:

> They were only clusters of four and six cells, but in them ACT's scientists saw a revolution in medicine that will render many of today's drugs

and treatments obsolete.... Unlike existing stem cell lines, stem cells created through cloning would provide a patient with a fresh supply of cells with his or her own genetic code. Gone would be transplant failures and the need for immune-suppressing drugs. In the same way that antibiotics and vaccines rid the world of infectious plagues a half century ago, says Lanza, these cells could for the first time eradicate the chronic degenerative diseases of our day, such as cancer, Alzheimer's, and heart disease.[70]

If this kind of journalism causes investors' mouths to water, Michael West would be the last person to discourage such Pavlovian reactions. *Business Week* reported in 2002 that he estimated "the potential market [for therapeutic cloning] to be well over $10 billion per year."[71]

Fischer might have been more guarded in her reportage if she had crosschecked with other scientists. It didn't take long for critics to begin the debunking. For example, Gary Stix reported in *Scientific American* that ACT's work was highly questionable:

> Many leading scientists ... say the work should never have been published. First, ACT didn't produce any stem cells. But more fundamentally, some investigators question the company's basic assertion about having actually cloned a human embryo because eggs can sometimes be stimulated to divide briefly without fertilization.[72]

Unlike Fischer, Stix looked into the impracticality of therapeutic cloning, a too often neglected part of the story. Harry Griffin, assistant director of the Roslin Institute (where Dolly was cloned) informed him that even if therapeutic cloning could be technically accomplished, it would be most unlikely to become a revolutionary medical breakthrough:

> The suggestion that the cloning of an embryo would revolutionize stem cell therapy by providing a route for routine immunocompatible cell transplants is simply naïve. Such a treatment might be possible for a small number of patients, but there are five million suffering from Parkinson's disease in the U.S. alone. There is, in our view, no way that individual embryos can be created to provide individual treatments for this number of people—it would be incredibly costly, and there are simply not enough eggs available.[73]

Rudolf Jaenisch, a cloning expert at MIT, was equally scathing about ACT's supposed cloning breakthrough, labeling the experiment "third-rate science."[74]

This reality check did nothing to stop the media puffery of Advanced Cell Technology's sensationalistic activities. Less than seven months after the *U.S. News* piece, the *Atlantic Monthly* published a sixteen-page article on the wondrous goings-on in ACT's labs. Titled "Cloning Trevor," the piece was the work of Kyla Dunn, identified by the magazine as a "former biotech researcher." The story concerns a desperate family whose two-year old son "Trevor" carries a "rare and devastating genetic disease."[75] The piece opens with the company's ace technician Cibelli and a few colleagues preparing to clone embryos using Trevor's skin cells. About to take "a first step toward creating radical new cures" for patients like Trevor, they listen anxiously to the radio as President Bush presents his 2002 State of the Union address. "In perhaps no other field of science are researchers as mindful of which way the political winds are blowing," Dunn accurately comments. The scientists worry that Bush will push Congress to outlaw all human cloning, which would shut down their experiments: "Politics and religion, it seemed, were trumping science."

The earlier experiment that had been extolled in such grandiose fashion by Joannie Fischer turns out not to have been such a big deal after all, being "largely judged preliminary, inconsequential, and certainly not worthy of headlines." But no matter: Cibelli measures progress "not in years but in eggs," expecting to get therapeutic cloning to work "about two hundred eggs from now." Toward this end, ACT has stored skin cells from five patients, "one with a spinal-cord injury, one with diabetes, two with healthy but aging bodies, and Trevor." We are not told whether these patients paid ACT for this service, or whether they were promised that the company would attempt to treat them with cloned ESCs if its scientists succeeded in creating cloned embryos from their DNA.

Finding sufficient human eggs to conduct cloning experiments is Cibelli's key problem of the moment. Egg cells are expensive. ACT, readers are told, recently paid $22,000 to a young woman to harvest just ten eggs. As a result, "Instead of working with a hundred embryos," Cibelli complains, "I'm working with one." After the success-that-really-wasn't-a-success was reported in *U.S. News,* things have not gone well for ACT, and now the company is running short of funds.

Trevor's rare genetic disorder, ALD (X-linked adrenoleuko-dystrophy), "can abruptly ravage the white matter of the brain, with devastating and often fatal results." The disease can be treated with bone marrow or umbilical-cord blood transplants from a well-matched donor, a procedure that "can halt or even reverse the progression of the disease"; it seems that "the healthy transplanted cells travel to the brain, where they are able to prevent further damage." But compatible donors may be hard to find, and sometimes the patient's immune system rejects the transplanted cells, causing a life-threatening condition. So, rather than risk this dangerous treatment, Trevor's parents travel to ACT seeking therapeutic cloning for their son.

They meet with Michael West—"a dreamy, laid-back forty-eight-year-old with a lopsided grin." He agrees to help, but (as far as we can tell from the story) doesn't tell them of the many insurmountable problems that would impede treating Trevor with cloned ESCs. Not only would cloned embryos have to be made, using his DNA, and then ESCs extracted—two very daunting tasks—but the cloned embryos would require gene modification to correct the ALD-causing defect. At the time of the story, scientists had not yet been able to create a cloned stem cell line or transform animal ESCs into bone marrow cells, as they planned to do for Trevor.

But West is upbeat. "This is doable," he assures the parents. "It's only a question of can we mobilize the people. It won't happen unless people work on it."

Following the usual media practice of putting down ASCs, Dunn quotes CEO West, who surely knew that potential investors would be reading the article, extolling cloning research over all other approaches:

> Where there's this ethics debate about adult versus embryonic stem cells and cloning. . . . I don't think what's properly weighed in the balance is the amazing breakthrough that this is. I mean, the idea that you can take a person of any age—a hundred and twenty years old—and take a skin cell from them and give them back their own cells that are young! Cells of any kind, with any kind of genetic modification! That's such an incredible gift to mankind! For the U.S. Congress to spend two hours and debate this and say, 'Oh, we'll make all this illegal,' to me is unbelievable. They don't understand."

Dr. Lanza waxes even more hyperbolic:

> I'd stake my life on it. If this research is allowed to proceed, by the time we grow old, this will be a routine thing. . . . You'll just go and get a skin

cell removed at the doctor's office, and they'll give you back a new organ or some new tissue—a new liver, a new kidney—and you'll be fixed. And it's not science fiction. This is very, very real.[76]

Naturally, no ASC researchers or opponents of cloning are allowed to pour cold water on the overheated sales pitch. The piece ends after Cibelli fails to clone Trevor, and early symptoms of ALD force his parents to turn to traditional treatments: "All hopes of developing an experimental cure for Trevor were dashed; time had run out."

In reality, there never had been any hope of treating Trevor with therapeutic cloning. It is not apparent from the information in the article that West made it clear to the parents that human clones had not yet been successfully created. Nor do they seem to have been told that it would have been unethical, and probably illegal—because of the dangers involved—to treat Trevor with cloned ESCs. In short, West appears to have used the child's tragedy and his parents' desperation so as to reap free publicity in a national magazine.

And the beat goes on: In January 2004, the high-tech magazine *Wired* made international headlines by reporting that ACT had successfully maintained cloned human embryos to the 16-cell stage. As with the *U.S. News* and *Atlantic Monthly* stories that had preceded it, *Wired*'s journalist was also granted exclusive behind-the-scenes access to the supposed genesis of a new epoch. And as before, the embedded writer was more propagandist than journalist. "I'm sitting beside Robert Lanza, medical director at Advanced Cell Technology," Wendy Goldman Rohm writes. "He's breathing softy, hands folded neatly in his hap, his head bowed as if in meditation. For years he's been preparing for this day—making plans, conducting preliminary tests, losing sleep. Now, on October 12, we're six hours into the experiment and all he can do is watch."[77]

ACT intends to manufacture cloned embryos after purchasing eighteen human eggs from two women for $8,000. After the *Atlantic Monthly* article, Jose Cibelli left ACT to take a position at Michigan State University—where, ironically, it is against the law to engage in human somatic cell nuclear transfer.[78] So this time it's a young scientist named Young Chung who performs the delicate laboratory tasks of SCNT under Lanza's watchful eye. "If it works," Rohm says, "Lanza will have accomplished two amazing things: He will have cloned human embryos, and he will have harvested stem cells from them. Embryonic stem cells are prized for their magical potential to become any type of cell in the body." Better yet, "Clone embryos produce stem cells that are exact genetic matches of the donors and consequently run little chance of being rejected."

Typical of this genre in the mainstream media, the technical account is inaccurate. As we noted earlier, cloned embryos can never be exact matches to the DNA donor unless she also provided the egg for SCNT. Nor have ESCs ever been extracted from human embryos at the 16-cell stage. And Rohm quotes Lanza's assertions that therapeutic cloning could provide "someone who has a heart attack" with "a ready supply of stem cells," but does not point out that human heart attack patients have *already* in early human trials received this clinical benefit from being treated with their own blood or bone marrow stem cells.

Helping with the fundraising, Rohm writes, "ACT's future depends on developing additional technologies that it can patent and license." CEO West lets on: "We'll need tens of millions to carry the research through to clinical trials and to the successful launch of the product."[79]

On the fifth day, Rohm is banished from the lab. "I have witnessed nearly every minute of the experiment," she writes. But Lanza's "increasing obsession with not disturbing the embryos" led to her exclusion.

Later, her phone rings.

"Are you sitting down?" Lanza asks.

"What have you done?"

"We're so surprised. We did it!"

Lanza tells her that they have achieved one 16-cell cloned human embryo. This would be big news if verified. But there is no way of knowing the truth. Lanza had not—and has not as of this writing—submitted his results to peer review nor published in a respected science journal. And despite *Wired*'s press release claiming that their reporter Rohm "witnessed" the experiment, she never saw firsthand proof that the human cloning attempt had actually succeeded.[80] "I've learned nothing more about the human clone embryo," she winds up, "and, as I return to my hotel, I am left pondering a host of questions. Did the embryo progress beyond the 16 cells to a blastocyst? Will it yield stem cells? Or has it already died? ... But this much is clear: ACT is pioneering new methods to grow stem cells—and along the way, bringing us closer to a fascinating, if ethically complex, future."[81]*

These starstruck articles—and the plethora of similar stories that have appeared regularly over the last few years—bear witness to the power of special interests to generate favorable publicity. The basic strategy: to keep on raising false hopes.

Rent an Ethicist

In addition to cozying up to reporters, some companies—anxious to justify research that many believe to be immoral and unwise—retain bioethics consultants to "advise" them on what should and should not be done. But this so-called safeguard against unethical conduct is actually more for show than anything else; there is little risk of an impartial ethical evaluation since most of these people wave the flag for the panoply of new technologies. Those who are known to be skeptics are unlikely to be put on the company payroll.

Of course, worries about the sincerity of bioethicists are nothing new. In 1997, Ruth Shalit wrote a seminal article about "ethics for hire" in the *New Republic:* "Increasingly," she wrote, "medical ethics consultants are being retained by HMOs to assist them in decisions about grievance cases, risk management, and 'non compliant' patients." The job of these consultants is to help justify executive decisions; those who raise awkward caveats or outright refuse will probably not be invited back.[82]

The same is true of the biotech field: many bioethicists have grown increasingly cozy with industry barons in recent years. Even some in the bioethics establishment have grown concerned about this trend. In "Throwing the Watchdog a Bone," which appeared in the prestigious bioethics journal *Hastings Center Report,* University of Minnesota philosophy professor Carl Elliot didn't mince words:

> Many bioethicists are no longer clearly scholars or clinicians, but a strange
> hybrid of policymaker, pundit, and bureaucrat, floating on the borders of
> the government, the business world, and advise industry. Unmoored from

---

*After the *Wired* story made world headlines, West refused to verify that ACT had created a 16-cell human cloned embryo. "Cloning and Stem Cells: Baby Steps," *Economist,* December 30, 2003, p. 53.

a tradition, unanswerable to a professional code, in the university but not quite of it, bioethics is fast becoming another cog in the complex machinery of business, entertainment, and politics that controls the shape and direction of American life.

In short, "each corporate check cashed takes us [bioethicists] one step closer to the notion of ethics as a commodity."[83]

Daniel Callahan, one of the originators of bioethics as an intellectual discipline, has likewise raised the alarm that the field to which he has devoted his professional life may be "succumbing to the lure of money." Asking if bioethicists are selling ethics, Callahan warns of "disturbing signs" that mercantile values are creeping in. "The greatest problem comes in working with the for-profit sector, getting paid for ethical advice to companies whose purpose is to provide a good return on investment to stakeholders."[84] But for those of his colleagues who want the money and don't regard their ethics counsel "as a one-time-only kamikaze mission," Callahan offers this advice:

> [Y]ou will not be retained as a consultant if you publish anything likely to hurt the image of the company you are serving—or unless you agree not to publish anything at all without their permission. . . . There is just no way of acting as a consultant-for-hire without running the risk of either alienating those who pay, or not alienating them and appearing to make them look good. If one takes the money, then one has to put oneself, and one's work, in their hands.[85]

These views from a bioethics pioneer may be the last nostalgic gasp of those who regard their subject not as a job, but as an intellectual and moral calling. As the biotech industry increasingly becomes aware of the tremendous public-relations benefits to be derived from the approval of so-called ethical authorities, checks are being written—and cashed. It was a sign of the times when an online trade publication recently discussed how the "business of bioethics" could promote biotech companies' bottom lines: "For an increasing number of companies and for bioethics consultants with them, this business is booming. Biotechs and pharmas [pharmaceutical companies] have, for some time now, begun consulting with experts in ethics, setting up bioethics advisory boards," activities that go "hand-in-hand . . . with public relations."[86]

Receiving substantial fees—or better yet, stock options—can be seductive inducements to lending one's bioethical expertise. Rahul K. Dhanda urges bioethicists to join in the applause for biotechnology's pursuit of "new knowledge ... harnessed in the service of capitalism."[87] Pointedly disagreeing with philosophers like Daniel Callahan who prefer impartial analysis, this corporate bioethicist conjures a different vision in which the discipline in question "becomes a seamless part of the corporate landscape."[88]

Dhanda's prime example of how bioethics can keep the industry on a moral path was a decision by the Geron Corporation's ethics advisory board: namely its "unanimous judgment that research on [ESCs] can be conducted ethically." Ludicrously, the board even issued a statement to Geron's CEO, Thomas Okarma, assuring him that the company's planned research into cloning and ESC research would be ethical, so long as Geron treated its embryos "with moral seriousness" and did not engage in "cloning for purposes of reproduction"—among other equally vague and easily complied-with directives.[89]

Of course, pronouncements like this are not only just predictable: they are unenforceable. Dhanda admits that companies are free to ignore the advice of their ethics advisory boards (EABs). An example of this was when University of Pennsylvania professor and bioethicist Glenn McGee (a supporter of cloning-for-biomedical-research) resigned from the Advanced Cell Technology EAB, accusing his colleagues of being "rubber stamps" after the company began cloning human and animal embryos without first obtaining committee approval.[90] McGee's integrity is admirable. But his public protest caused scarcely a murmur and didn't affect the company's research program one whit.

Dhanda comments that Geron's ethics board had no direct financial interest in the company—which is not the same thing as saying they weren't paid. He dismisses Callahan's contention that bioethics advisors need to remain aloof from financial ties to the corporations they advise. "Some might argue that, ideally, the EAB should have no financial stake in the company," he writes. But this as an "unfounded concern.... These individuals are providing a substantial service to the company, and if the EAB shares the corporation's vision that a bioethical company leads to a successful company, providing a stake in the company may be an incentive to the EAB members to better perform the responsibilities of their office."[91]

So this is the future: bioethicists serving the captains of biotechnology. For both parties, it could be the perfect match. As for the rest of us, we should be aware, as William Saletan wrote in *Slate,* that "corporate ethicists, like corporate lawyers, have reduced their purview to technique. Tell them what you want to do, and they'll tell you how to do it."[92] Or to put it another way: when ethics advisory boards see fit to justify morally problematic research, we should "consider the source."

Corporate Welfare for Big Biotech

Private investors have poured almost $100 billion into biotech in the last quarter-century, according to the *Wall Street Journal.*[93] In 2001 alone, venture capitalists worldwide invested a whopping $10 billion into the business. Pharmaceutical conglomerates have also added to the industry's financial clout by collaborating with biotech startups, partnering with them or buying them outright. According to a report by the accountancy firm Ernst & Young, there were 480 "pharma-biotech" collaborations worldwide in 2001.[94] To give an idea of the cash available to these companies, the Biotechnology Industry Organization, the industry's American lobbying arm, spent $142.9 million between 1998 and 2002, with $7.7 million in 2002 campaign donations alone.[95]

Investment statistics are not the same thing as being profitable, of course. Biotechnology is a relatively new industry and its expenses are enormous: overall it has lost some $40 billion.[96] So it's hardly surprising that many companies have yet to provide a return for their investors.

And some areas of the industry are doing less well than others. The rule is: The more controversial, the more financially strapped. Advanced Cell Technology, which has led the charge on human cloning-for-biomedical-research, is so chronically low on cash that its researchers have reportedly financed some of their own work. Geron's stock plummeted from a high of $75 per share to around $10, with attendant layoffs, in the spring of 2004.[97] Other companies seeking to create products from human cloning or embryonic stem cell research, such as PPL Therapeutics (UK) and Pharming (the Netherlands), have also had significant financial difficulties. The *Financial Times* sums up the situation: "The finances of the world's cloning companies are so precarious that a lack of funding may accomplish what moral objections have so far been unable to do: bring research in the area to a halt."[98]

There are abundant reasons for caution on the part of investors:

- Even if human cloning and/or ESC therapy can be reliably accomplished—no sure thing—it will take many years for the technology to become usable as a medical technique. In the meantime, cloning research is a black hole that sucks investments in, but releases no returns.

- Even if biotechnologists succeed in creating reliable human cloning techniques and ESC therapies, the companies will have to obtain patents and the approval of the FDA before attempting therapeutic cloning in humans. Given the dire safety problems (such as tumor formation associated with ESCs in animal studies) that cloning procedures would not remedy, this project would probably take many years and hundreds of millions of dollars to accomplish.*

- Human cloning is explosively controversial and cloning opponents remain adamant that human SCNT be banned. Even though this effort has not yet succeeded in the United States, it has in other countries. This reality may have contributed to the difficulties that U.S. companies are having in obtaining investments. "Some executives and analysts say the controversy has kept companies and investors from the field," the *New York Times* reported, while the "attention [paid to] the potential of stem cells has spurred investment, *particularly in companies using non-embryonic cells.*"[99] Michael West, head of ACT, the company pursuing human cloning, was quoted by the *Atlantic Monthly* in 2002: "Raising money is difficult when Congress is trying to criminalize your business."[100]

- The omnibus spending bill that passed Congress in January 2004 outlaws the patenting of human life for at least a year. Once such provisions get into omnibus bills, they tend to stay

---

*However, Geron claims to be close to seeking permission to begin early human trials with ESCs in the treatment of recent spinal injuries. *Daily Reproductive Health Report,* March 17, 2004. Company officials claim that if they obtain permission, their finances will turn around. Luke Timmerman, "Stem-Cell Research Is Exciting, but Not to Investors," *Miami Herald,* March 20, 2004, reprinting an article that originally appeared in the *Seattle Times.*

there indefinitely. Without patents, therapeutic cloning cannot proceed profitably.

- Adult stem cell and associated regenerative medical approaches are moving forward at a tremendous pace, spurring heavy private investment in a field where "the practical use of adult stem cells is not 10–15 years away but well along in the commercialization process."[101] If this pattern holds, companies pursuing these uncontroversial technologies may soon begin to turn profits, leaving their more controversial competitors in the dust.

Reporter Luke Timmerman concluded that investors "aren't committing billions of dollars," because "society hasn't clearly decided whether the research is moral, the field is too risky, the business model too vague. Researchers don't know how to control embryonic stem cells ... and they don't know how to do it cheaply, conveniently, or consistently enough to make it a viable business."[102]

Facing a financial black hole, Big Biotech and its allies have decided to turn to Plan B: gaining access to public coffers so that taxpayers foot the bill. But that's not as easy to accomplish as they would like. Human cloning is an unpopular technology. For example, a 2002 Gallup poll found that the American people opposed "cloning embryos for use in medical research" by 61–34 percent.[103]

To overcome public resistance to federal financial support for research that most people oppose, Big Biotech and its apologists mounted a three-pronged public-relations onslaught and a parallel political campaign:

Blur the distinction between ESC research and somatic cell nuclear transfer (SCNT). The reader may recall how in the spring of 2001, President Bush was pressured by Big Biotech to fully fund ESC research. At the time, the promoters assured a leery nation that all they wanted was funding for research using embryos, "in excess of need," from IVF infertility treatments—that is, embryos who were going to be discarded anyway. *Never,* they solemnly assured us, would they countenance the creation of human embryos solely for research.

Today, the party line is that SCNT, when not undertaken to bring a cloned baby to birth, *is ESC research.*[104]

Pass state laws defining SCNT as a form of ESC research. As we've seen, ESC research and SCNT have already been closely linked in a few states. In 2002, California passed a law explicitly authorizing SCNT as part of a broad ESC research authorization. New Jersey followed suit, in 2004 enacting a law so radical that it permits the gestation of cloned human fetuses through the ninth month. Similar bills were introduced in Texas, Maryland and Delaware.

The strategy here is obvious: Once a handful of states pass such Big Biotech-spawned laws, they could then be touted as expressing the "will of the people." Moreover, cloning advocates could argue that banning SCNT at the federal level would interfere with states' rights. This bottom-up approach might even pave the way for the passage of the Hatch/Feinstein phony ban, or failing that, be used as justification for an executive order permitting federal funding of SCNT so as to level the playing field throughout the nation.

Obtain state funding pending a change in administration. Within a week after New Jersey's cloning license went into effect, biotech boosters began complaining about a lack of state funding. As a partial curative, Ira Black, chairman of the Department of Neurosciences at Robert Wood Johnson Medical School, advocated in the *Trenton Times* that funds from "existing state commissions, such as the one dedicated to spinal-cord research," be diverted to pay for experimental ESC research and SCNT.[105] More recently, the governor of New Jersey called on the state legislature to appropriate millions of dollars each year to help finance embryonic stem cell and cloning research.

California has authorized state funding for SCNT research. Given the state's catastrophic budget deficit, however, it is unlikely that taxpayers will do so. In a truly audacious move, Big Biotech campaigners launched an initiative to appear on the November 2004 ballot, backed by a reported $20 million budget, to persuade California's voters to pass a constitutional amendment that would permit ESC research and human SCNT. The initiative would also establish "The California Institute for Regenerative Medicine," a sort of miniature NIH tasked with a pro-biotech mission. Topping it all off, the initiative would pay for it all by borrowing $3 billion over ten years via a bond measure that would require Californians to pay biotech companies and universities an average of $295 million per year, with "priority" given to the controversial technologies.[106] The ultimate cost of the measure: $6 billion over 30 years, $3 billion in principal and $3 billion in interest payments.[107]

■ ■ ■

There is no doubt that most people who support human cloning-for-biomedical-research earnestly desire to help ill and disabled people live better lives. But this is not the only motive for their advocacy. Explicit or implicit scientism drives many of them to fervently oppose *any* prohibitions on scientific research based on the moral values of society—particularly if they believe these to be founded in religion. Add the quest for riches through biotech entrepreneurship into the equation, backed as a fallback position with tax dollars, and the stage is set for the emergence of a dehumanized future.

CHAPTER 6

Will Humans Remain Human?

· ·

T IS A NATURAL HUMAN DESIRE to manipulate our bodies to look better, feel better and age better. We not only wish to be free of disease, but also deeply desire to remain youthful in appearance and physical vigor. "Wellness" isn't just about enjoying "good health" anymore. Indeed, for some of us our bodies aren't just bodies, they are the canvasses upon which we paint our individuality.

We also use "artificial" means to enhance our looks and improve our bodies and performance. Young people alter their appearance through the ancient practices of tattoo and body-piercing. Older people attempt to reverse the appearance of aging with "Botox parties" and cosmetic surgery. Drug companies pitch virility pills on television to convince middle-aged men that they can regain the libidos they had at age eighteen.

Supporters of human biological enhancement would say that harnessing the power of genetics to accomplish similar goals is not different in principle from the dowager's facelift. And is the dad-to-be who purchases a biotech body design that inserts a "petite gene" into an embryonic daughter to improve her chances of becoming a great gymnast, they might ask, doing something very different from Tiger Woods's father when he developed his son's talent from a very early age?

But such thinking confuses ends with means. It ignores the radical difference between superficial bodily enhancement and deep biological alteration, and is blind to the ramifications of genetic tampering for self, progeny and society. Suppose I got a facelift, got my mostly silver hair dyed jet-black, had a dentist create a new, handsome smile with straighter and whiter teeth, and got an implant so that my chest looks like the young Arnold Schwarzenegger's. Given sufficient funds, I could so alter my appearance that my own mother might not recognize me. But these changes would remain, quite literally, skin-deep: my biological nature would remain unchanged. And if I had children, the genes I would pass on would of course remain unaltered.

But what if we could fulfill our dreams for ourselves or for our children through permanent genetic or chromosomal manipulation? What if I could genetically enhance myself so that my hair would never turned gray and my body never lose the leanness and muscle tone of youth? What if a father could insert a gene to transform his daughter into the concert pianist *he* always wanted to be, or an atheist do likewise to ensure that his children would be genetically predisposed (if it proves possible) to shun religious belief? And what if these modifications passed down the generations?

Questions such as these are no longer merely a good subject for student papers in a college moral philosophy class. They are quickly becoming crucial moral and political questions that society must debate and answer. For with the coming of Dolly, the mapping of the human genome, and now the cloning of human embryos, we are forced to come to grips with the potential that biotechnology will be able to change the future evolution of the human race.

The Door Opens to Brave New World

On October 5, 2000, a biotech consortium announced a research "breakthrough" that sent Joseph Bottum, the book and arts editor of the *Weekly Standard,* into high apoplexy. "It was revealed," he wrote, "that biotechnology researchers had successfully created a hybrid of a human being and a pig. A man-pig. A pig-man. The reality is so unspeakable, the words themselves don't want to go together."[1]

The SCNT cloning experiment, conducted by the Australian company Stem Cell Sciences and the American company Biotransplant, made a pig-human hybrid embryo by inserting a human somatic fetal cell into the egg of a pig; the patent application claimed to have developed it to the 32-cell stage. Adding to the intensity of the controversy, Peter Mountford, head of Stem Cell Sciences, made the dubious assertion that the embryos could theoretically have been implanted either in a woman or in a sow. Some observers even suggested that mixing species was an important step toward enhancing the human genome and creating what visionaries sometimes call a *posthuman* race.

"You can't say we weren't warned," Bottum admonished, before launching into hyperbole:

> This is the island of Dr. Moreau. This is the brave new world. This is Dr. Frankenstein's chamber. This is Dr. Jekyll's room. This is Satan's Pande-monium, the city of self-destruction the rebel angels wrought in their all-consuming pride.... In the midst of this, the creation of a human-pig arrives like a thing expected. We have reached the logical end, at last. We have become the people that, once upon a time, our ancestors used fairy tales to warn their children against—and we will reap exactly the conse-quences those tales foretold.... You and I will live for many years in youth-ful health: Our cancers, our sensibilities, our coughs, and our infirmities all swept away on the triumphant, cresting wave of science. But our sons and our daughters will mate with the pig-men, if the pig-men will have them. And our swine-snouted grandchildren—the fruit not of our loins, but of our arrogance and our bright test tubes—will use the story of our generation to teach a moral lesson to their frightened litters.[2]

Some public intellectuals share Bottum's dread of a sinister antihu-man outcome. But others are elated, fervently looking forward to the bright new day of human redesign. The futuristic movement known as transhu-manism, for example, is organizing with the intensity of a religious revival.

While transhumanism is relatively new, the idea that we should apply the full array of new technologies to remake the natural human order has been bubbling up in radical bioethics and academic philo-sophical discourse for decades. We need only return to the views of Joseph Fletcher, the "patriarch of bioethics," for proof. Professor Fletcher was a devoted believer in an anything-goes approach to Brave New World innovations.[3] He believed that nothing in the natural way of life was sacrosanct, and he promoted a social "revolution" that would use arti-ficially induced mutations to usher in a "quantum or dialectic leap in human change ... in the minds and flesh of human beings," toward the goal of creating "superior people."[4]

Much of Fletcher's preaching—with his call for human/animal chimeras to become our menial laborers, for instance—could be taken right out of *Brave New World*. As an example of the unnatural possibil-ities he prophesied as our destiny, Fletcher swooned at the prospect of males giving birth:

[T]ransplant or replacement medicine foresees the day, after the automatic rejection of alien tissue is overcome, when a uterus can be implanted
in a human male's body—his abdomen has spaces—and gestation started
by artificial fertilization and egg transfer. Hypogonadism could be used
to stimulate milk from the man's rudimentary breasts—men too have
mammary glands. If surgery could not construct a cervical canal the deliv-
ery could be effected by a Caesarean section and the male or transsexu-
alized mother could nurse his own baby.[5]

A remarkable fantasy, no question. For most people, the very notion of
a man-mother is nightmarish. But for those who prize radical autonomy
above all else, accepting the limits of our nature is not wisdom but
oppression.

Fletcher's views are not aberrant or even on the fringe of modern
bioethical thinking—though few would advocate human/ape copula-
tion. Rather, the predominant view in contemporary bioethics about
these matters can be summarized in that word that trumps all others:
Choice.

But there is more to this story than fulfilling personal desires. In
The Lives to Come: The Genetic Revolution and Human Possibilities, Philip
Kitcher admits that eugenics thinking has returned. But that's okay, he
believes, since it will be a "laissez-faire eugenics" in which people will
create their own versions of optimal human life—a prospect that Kitcher
naïvely assures us will work out just fine because there will be a "universally shared respect for difference."[6]

DNA double helix co-discoverer James Watson has urged that
prospective parents be allowed to eradicate undesirable traits—or introduce desired enhancements—in their children-to-be through the wonders of genetic engineering. "Who wants an ugly baby?" Watson asked
in a lecture. "Going for perfection was something I always thought you
should do. We'll be able to make correlations between genes and certain professions, genes for the undertaker—they really don't cry very
much—or the sprinter. It will be an absolute flood. . . ."[7]

Some believe that we should bioengineer our progeny so as to prevent *other* nations from dominating the world by creating superhumans
with vastly improved capabilities. Political pundit Tony Blankley, for
example, currently the *Washington Times* editorial page editor and someone who would otherwise be opposed to treating human life as a malleable commodity, reacted hysterically to the announcement by a Chinese

bioscientist that she had created cloned blastocysts using human DNA and rabbit eggs. Fearing that the Chinese were beginning a massive research program that would result in the creation of a "group of super-intelligent people" and hence provide them with a "strategic geopolitical advantage," Blankley urged Americans to abandon their moral qualms and jump immediately into a bizarre genetic arms race:

> Bioengineering may be able to physically [and intellectually] improve man across the entire spectrum of our functions—yielding extraordinary economic as well as strategic advantages. (Neanderthal man was a magnificently successful early man. But when he met the more intelligent CroMagnon man, he quickly went extinct.)
>
> Of course, mistakes will be made. Island of Dr. Moreau-like monsters may well be formed. God may punish a people who presume to tinker with his handiwork. But as the Chinese push forward, hell-bent for industrial levels of genetic manipulation and cloning, supported by the massive bioengineering research they are now beginning to fund, American voters and congressmen will have to balance their strong ethical and religious revulsion of cloning against the danger of being surpassed by a gene-manipulated super-race.[8]

Bioethicist Gregory E. Pence promotes human reproductive cloning as the most effective route to genetic improvement. Why cloning? The strongest argument for permitting cloning-to-produce-children, he wrote in *Who's Afraid of Human Cloning?*, is that gifted parents could by this means confer on their children a "wonderful genetic legacy."[9]

The deeper one delves into the posthuman agenda, the clearer it becomes that dissatisfaction with natural humanity lies at its heart. Consider, for example, Gregory Stock's robust and enthusiastic support for creating a posthuman future in *Redesigning Humans*.[10] Stock, the director of the Program on Medicine, Technology and Society at the UCLA School of Medicine, believes that all individuals should be absolutely free to alter themselves and their progeny. This could include inserting animal DNA into human embryos, inserting or removing chromosomes, inserting artificial chromosomes into a genetically engineered embryo, or perhaps altering human capacities through nanotechnology.

Stock envisions a time when we will become so genetically diverse that humans will cease to be a single species. We may no longer be able to procreate outside of the laboratory since the "union of egg and sperm from two [transhumanist] individuals with different numbers of chromosomes or different sequences of genes on their extra chromosomes would be too unpredictable . . . ." If that sounds as if having children will become onerous, don't worry. Stock predicts that our attitude toward children as tailor-made commodities will have become so ingrained that laboratory conception will not "seem a burden because . . . parents will probably want the most up-do-date chromosome enhancements anyway."[11]

Such thinking is echoed by Lee M. Silver, a Princeton biology professor whose *Remaking Eden* is the most famous and influential apologia for human cloning and genetic engineering published to date. Silver's fervor for human cloning is based on his belief that this technology will be the key that opens the door to a redesigned human condition: "Without cloning, genetic engineering is simply science fiction. But with cloning, genetic engineering moves into the realm of reality."[12]

Silver points out that genetic engineering—already being accomplished in animals—is very "inefficient," with a success rate of 50 percent at best, and with the additional risk of adverse genetic manipulations creating new strains of disease. This is not a problem for geneticists working with animals, Silver says, since they can pick out healthy specimens and destroy the defective ones—hardly an option when it comes to human lives (to which the modifier "yet" must be added).

This is where cloning enters the picture. Cells will be extracted from a donor and the DNA in the nucleus genetically engineered to taste, and then used in human SCNT cloning. Once the cloned embryo reaches the blastocyst stage—five to seven days of development—it is implanted into a woman's womb and gestated to birth. The child's genes will be virtually identical in genetic makeup to the modified cell from which he received almost his entire DNA. In theory, this would result in the child exhibiting the desired "enhancements."

Once the technology becomes widely accessible, Silver predicts, "the global marketplace will reign supreme," resulting in a genetic arms race in which the affluent will compete with each other to enhance their children with increasingly sophisticated modifications. In addition to the obvious general improvements that he believes could be made—in health, strength, beauty and intelligence—he doesn't balk at the idea of introducing specialized "animal attributes . . . into the human genome": to increase our sense of smell, for example, or even to provide us with "light emitting organs" by means of firefly genes.[13]

In his own reprise of Huxley, Silver says that over time this competition will lead to transformations so radical that humanity will divide into two categories: the "Naturals" doomed to go through life unenhanced, and superior beings absurdly called the "GenRich." In an updated version of eugenic "master race" thinking, the *übermenschen* GenRich will utterly dominate the *untermenschen* Naturals:

> All aspects of the economy, media, the entertainment industry, and the knowledge industry are controlled by members of the GenRich class. GenRich parents can afford to send their children to private schools rich in resources required for them to take advantage of their enhanced genetic potential. In contrast, Naturals work as low-paid service providers or as laborers, and their children go to public schools.... Now, Natural children are only taught the basic skills they need to perform the kinds of tasks they'll encounter in the jobs available to members of their class.[14]

In the far distant future, the GenRich and the Naturals will become two entirely separate species. "In this era," sighs Silver ecstatically, "there exists a special group of mental beings" who "can trace their ancestry back directly to homo sapiens," but who are as "different from humans as humans are from the primitive worms with tiny brains that first crawled along the earth's surface."[15]

Such fantastical thinking has spawned the explicitly eugenic transhumanist movement. Promoted by professors from major universities and futurist fantasizers connected to each other via the Internet, transhumanism is steeped in materialism and an absolutist devotion to personal autonomy; it advocates seizing control of human evolution through bioengineering, species mixing, nanotechnology and other emerging techniques. The ultimate aim is to create a "posthuman" species, defined as "a human descendant who has been augmented to such a degree as to be no longer a human."[16]

Transhumanists are organizing to promote their posthuman agenda. They have formed international associations, they publish academic journals and online magazines, and they periodically gather at symposia to exchange views and discuss tactics. Perhaps the most prominent such group is the World Transhumanist Association (WTA), founded in 1998, dedicated to promoting the right of people to redesign themselves and their progeny so as to have "better minds, better bodies and better lives"— to be "better than well."[17]

In essence, transhumanism is a futuristic misanthropy. These people and kindred would-be enhancers think that human life has no special meaning in itself, but that the value of any life—animal, human, posthuman, machine, space alien—depends on the individual's measurable capacities, particularly his or her level of cognition. James Hughes, a bioethicist and professor at Hartford's Trinity College and one of the movement's most vehement polemicists, holds that rights are contingent on an organism's "level of consciousness." This "consciousness-based ethical standard" entails that posthumans will one day perceive themselves to be superior to "mere" humans.[18]

Transhumanists embrace extreme materialism and scientism. Driven by an ethos of radical individualism that countenances no restraints and disdains moral limits on personal behavior, believing that they possess the wisdom to improve the human species, longing desperately for corporeal immortality, transhumanists expect to mount a rebellion against nature that will, in the movement's eschatology, result in the literal recreation of human life. Thus Simon Smith, editor-in-chief of the transhumanist online publication *Betterhumans Newsletter,* writes: "Trans-humanists unabashedly assert that without gods, it is up to humanity to 'play God,' striving to achieve for humanity a total control over its physical and mental state, in some ways similar to that promised in supernatural beliefs."[19]

The ironic parallel between religion and transhumanism has also been noticed by Leigh Turner, a member of the medical faculty at McGill University, who has pointed out that transhumanists have substituted biotechnology for God as a higher power to worship. In "Biotechnology as Religion," published by the science journal *Nature Biotechnology,* she suggests that for "transhumanists, posthumanists, and other technological enthusiasts," biotechnology "provides a surrogate religious narrative" offering "many of the insights and rewards offered by traditional religious cosmologies." Seeing a close parallel to the eschatology of evangelical Christianity, Turner notes:

> Biotech, in a similar manner to many religious movements, has its charismatic prophets, enthusiastic evangelists, and enrapt audiences. Like religions, it offers a comforting message of salvation. Instead of imagining a day of rapture when the dead rise from their graves to begin eternal life, biotech enthusiasts imagine the era when medical technologies provide a renewable, largely imperishable body.[20]

In the transhumanist catechism, natural humans aren't good enough, smart enough, strong enough, athletic enough, beautiful enough, or genetically diverse enough. Transhumanists throw out the very idea of "normality," indeed of there being any virtue at all in remaining fully human. They fervently pray (in the secular sense, naturally) that anyone dissatisfied with their natural condition may be able one day to visit the local redesign clinic to have themselves or their progeny remade to order: as a pig-man, an orangutan-man, a part human/part cyborg, a woman remade to look like a cat, a supergenius, the strongest man who ever lived, or any other combination of human/animal/robot/computer, nanotechnology, synthetic blending of genes or surgical alterations that their polymorphous imaginations can come up with.[21]

This isn't just alarming, it is profoundly sad. Since transhumanists seem motivated by an existential despair over the limitations inherent in the human condition, it's rather paradoxical that one goal of the movement is physical immortality. This ambivalence reminds me of the old borscht-belt joke about a woman who asks her friend how she liked the town's new restaurant. "Awful!" the friend replies. "The food is terrible—and they serve such small portions."

The Transhumanist Declaration, published by the WTA, proclaims: "Transhumanists advocate the moral right for those who wish to use technology to extend their mental and physical capacities and to improve their control over their own lives. We seek personal growth beyond our current biological limitations."[22] Spouting notions that might have come out of the movie *The Matrix,* they apparently seek to achieve "The Singularity," a term referring to "the moment when technologies meld and an exponentially advancing intelligence is unleashed."[23]

Transhumanists enjoy casting themselves as morally superior to their opponents. Movement propagandist James Hughes, for example, castigates those who oppose the transhumanist pursuit of the unnatural as "cultural conservatives, nationalists, ethnic chauvinists or racists, religious conservatives, and opponents of women's equality, sexual freedom, and civil liberties."[24] But progressives—"secular, educated, and cosmopolitan"—will "take control of the natural and social forces that control their lives" and set up a world government whose most important responsibility will be to guarantee the universal right to transhumanize.

This would include gestating fetuses out of the womb and "mixing species boundaries through transgenic technologies, granting rights to the 'cryo-suspended' [people who have had themselves frozen], full nano-replication of the mental process, full identity cloning [persons multiplying themselves into new media], distributing one's identity over multiple platforms"—e.g., having one's consciousness downloaded into a computer or another person's brain—controlling the genetics of our progeny and "merging ... several individuals into one identity."[25]

In Hughes's vivid imagination, "democratic transhumanism" and "morphological freedom" will transform the world into a radically individualistic and utterly materialist utopia, in which pure egalitarianism will arise from the obliteration of all norms. With a triumphant war whoop, he bellows: "Let the ruling classes and Luddites tremble at a democratic transhumanist revolution. Would-be genemods [sic] and cyborgs! you have nothing to lose but your human bodies, and longer lives and bigger brains to win! TRANSHUMANISTS OF ALL COUNTRIES, UNITE!"[26]

This program for our future depends on destroying our belief in the uniqueness and importance of human life. As Leon Kass has noted, "In order to justify ongoing [cloning and genetic] research, intellectuals [are] willing to shed not only traditional religious views but any view of human distinctiveness and special dignity, their own included."[27] Meanwhile, other philosophers and bioethicists are questioning the very concept of humanity as a distinct species, preparing the ground for the eventual cloning and genetic engineering experiments that would attempt to give birth to human/animal hybrid creatures.

An article published in the summer 2003 *American Journal of Bioethics* epitomizes the approach. In it, Jason Scott Robert and Françoise Baylis of Dalhousie University contend that since, with the exception of identical twins, "every human genome is different from every other," and since "much of 'our' DNA is shared with a huge variety of apparently distantly related creatures (e.g., yeast, worms, mice)," then "the unique identity of the human species cannot be established through genetic or genomic means."[28] For these and other reasons, the authors conclude, "no extant species concept justifies the erection of the fixed boundaries between human beings and nonhumans that are required to make breaching these boundaries morally problematic."[29]

What might be the consequences of such thinking? According to A. M. Chakrabarty of the University of Illinois College of Medicine, "If species cannot be defined, then the fear of crossing the evolutionary boundary is irrational." Thus, lawyers should begin to work out the legal ramifications of the human-animal blending that will soon be upon us. "If we are to know whether we should euthanize some of these clones at or after birth, as has been done with many cloned animals, we should go beyond present philosophical discussions on the propriety of human or hybrid human-nonhuman cloning and discuss the legal consequences of such efforts."[30]

In just a few short years, these ideas have already progressed to the point that they may be influencing government decision-makers. A 2002 governmental report issued by the National Science Foundation and the U.S. Department of Commerce recommends that the government spend billions of dollars pursuing some of the very technologies that transhumanists, in their morphological quests, are craving. The report ("Converging Technologies for Improving Human Performance") urges that we mesh cutting-edge nanotechnology, biotechnology, information technology and cognitive science to develop methods that "enhance human abilities and efficiencies."[31]

Warning without a hint of irony that success "is essential to the future of humanity,"[32] the report claims that if only we pursue the dream and invest the billions of taxpayer dollars required for its fulfillment, we could eradicate poverty, end suffering and put an end to war: "The twenty-first century could end in world peace, universal prosperity, and evolution to a higher level of compassion and accomplishment.... [H]umanity would become like a single, distributed and interconnected 'brain' based in new core pathways of society." The potential end result: "Evolution [of humanity] transcending human cell, body, and brain."[33]

For all of their utopian dreaming, however, the transhumanists may have it backwards. The striving for conformity, rather than individuality, is often the stronger human drive. Mass alteration of our genomes might not lead to a vast smorgasbord of genetically unique beings, but to a stultifying, banal sameness.

Many of us, of course, have always followed the dictates of fashion-setters and entertainment celebrities in our choice of clothes, hairstyles and jewelry. But today, the phenomenal spread of cosmetic surgery, frequently commented on by the media, is an extraordinary manifestation of our preference for conformity. As author Catherine Orenstein noted in the *New York Times*, millions of us switch to television "reality" shows to voyeuristically feed our unappeasable appetites for physical perfection by watching average-looking people be transformed by radical cosmetic surgery or dentistry. One such program even depicts "men and women enduring radical reconstruction to look like their favorite movie stars." The results are "badly and blandly" remade people of reduced individuality:

> Because they undergo many of the same cosmetic procedures—breast and chin implants, nose and teeth-straightening, liposuction and hair-lightening—executed by the same surgeons and beauticians, the contestants on these shows ended up looking eerily alike. And not incidentally, like the two blondes who vied for the Bachelor's [another reality show] hand in marriage, who in turn, looked like Britney Spears. All could be knockoffs of the blond Nichole Kidman in the *Stepford* [*Wives*] movie posters.[34]

What if the day arrived when we could use genetic enhancements—and not just fashion or surgery choices—to follow the herd? Would we become a society made up of people with ten different body types out of an old *Twilight Zone* episode?

The Human Enhancement Agenda: Science Fact or Science Fantasy?

Before we prepare the obituary for natural humankind, however, we should take a clear-eyed look at whether a "posthumanized" future is a realistic prospect. Is the transhumanist project at all plausible? The most likely answer: not in the near term, or even the medium term. At least for the next few decades, these ideologues' reach in the academy and elsewhere will far exceed their scientific grasp. In the longer view, unfortunately, it may not always be so.

Let's start with a basic question: Can the human genome be genetically engineered? The answer, theoretically, is Yes. Animals and plants are already being genetically redesigned, using recombinant-DNA techniques. For example, female goat embryos—in what science writer Michael Fumento calls "one of the strangest projects in biotechdom"—were genetically adjusted with spider genes so that spider silk would be produced by adult ewes in their milk.* Meanwhile, the brain of a bioengineered mouse developed at Stanford University contains millions

---

*Spider silk protein is actually one of the lightest and strongest materials known to man. "A woven cable as thick as your thumb can bear the weight of a 747 airliner," Fumento points out; the implications of this striking fact have not gone unnoticed by

of human neural cells.[35] And genes were introduced into cows so that the animals were only 99.9 percent bovine—the remaining 0.1 percent being of human origin.[36] Cloned pigs were also developed to contain certain human genes, this time to learn whether genetically engineered herds could someday become organ suppliers in human transplant medicine.[37]

The primary reason Ian Wilmut and Keith Campbell created Dolly in the first place was to learn how to use cloning techniques to genetically engineer animals that produce substances useful in making medicines—a process called "pharming." Indeed, genetically altered animals are becoming so common that scientists worry about what would happen if they escaped into the wild and contaminated natural gene pools.[38]

Animals or plants that have genetic contributions from other species are generally known as "transgenic." Transgenic animals now proliferate in biotech labs around the world, and cloning is at the heart of the enterprise. As Wilmut and Campbell wrote:

> Polly [a transgenic cloned sheep created shortly after Dolly] illustrates why cloning and genetic transformation make natural bedfellows: why, indeed, cloning technology is needed if genetic engineering is ever to be an exact science and a routine technology (whether indeed, transgenesis will ever truly justify the title "engineering"). Transformation is best brought about when there are many cells to play with, and they are spread out in a dish; after [genetic] transformation they can be cultured further and assayed to see if they contain the required genes, before the cells are made into embryos [via cloning]. Generally, a transformed animal would subsequently reproduce by normal sexual means, to produce an entire lineage of normally reproducing animals that happen to contain the novel required gene.[39]

This technique could, of course, also be applied to humans should cloning be perfected and researchers learn how to modify embryos reliably. Medical science already engages in human genetic selection to prevent babies with diseases or disabilities from being born. This is done mostly via prenatal screening, followed in some cases by the eugenic

---

scientists and engineers. Michael Fumento, *BioEvolution: How Biotechnology Is Changing Our World* (San Francisco: Encounter Books, 2003), pp. 37–39.

abortion of fetuses carrying "undesired" traits such as Down's syndrome. But reproductive medical specialists also use pre-implantation screening and selection of embryos during IVF procedures to weed out defects, such as the single genes that cause Huntington's disease, cystic fibrosis and Tay-Sachs disease.*

Sex selection is becoming relatively common via sperm sorting, embryo selection and abortion. In fact, in India and China the natural balance between males and females has been thrown completely out of whack because so many prospective parents abort girls and try again for boys. Some hope that in the future these same processes could be used to "select in" desired traits, like greater height or increased intelligence.[40] Such "value preference" eugenics could grow in popularity if embryos or fetuses could be screened for genetic predispositions. A disturbing poll taken some years ago found that 11 percent of couples would abort if they learned that their unborn child was genetically predisposed to obesity.[41]

Will scientists soon learn how to move beyond "therapeutic" genetic screening for disease prevention, to using biotechnology to "enhance" a future child's genetic makeup with the aim of creating a "better than normal" child?[42] The good news—at least for those who do not share the transhumanists' longings—is that "designer babies" will not be engineered anytime soon. According to a thorough study on the matter undertaken by the President's Council on Bioethics, "dreams of fully designed babies, based on directed genetic change, are for the foreseeable future pure fantasies."[43]

The council noted that most traits the transhumanists might wish to enhance—such as intelligence, appearance, memory, athletic ability and sensory capacities—involve very complex interactions among several genes that turn on and off and otherwise interact in ways we are nowhere near understanding. As if that weren't daunting enough, what we eat, how we are treated, our experiences from the womb into active life—injuries, friend-

---

*The first live birth of a baby chosen as an embryo for implantation after genetic screening for CF was a girl named Chloe, born in 1992. During IVF, scientists destroyed embryos with two CF genes, one from each parent, and implanted only those embryos that would not develop the disease after being born, resulting in Chloe's birth. Jeremy Rifkin, *The Biotech Century* (New York: Jeremy P. Tarcher/Putnam, 1999), pp. 134–35.

ships and other unique environmental factors—also affect our abilities, personalities, intelligence and looks.[44] Evidently the young Adolf Hitler was not predestined by his genes to become what he became; his individuality resulted from interactions of such complexity that it would be easier (despite the *Boys from Brazil* scenario) to play a dozen identical games of Scrabble in a row than it would be to manufacture a new *Führer.*

Even if the desired trait involved only a few genes—for example, eye color, height, skin color, perhaps even lifespan—"there would be no guarantee" that the "predisposing genes, even if correctly and safely introduced into the zygote or early embryo, would necessarily express themselves as desired, to yield the sought-for improvement."[45] All in all, then, dreams of a posthuman race are likely to remain just that.

Still, just because it can't be done today, next year or next decade, we should not be complacent about the threat to human dignity and equality posed by the quest for human enhancement. Although the genetic sciences are still in their infancy, we have already mapped the human genome and are now about the task of studying the complex ways in which genes express and interact with each other, aided by increasingly sophisticated computer programs that are exponentially increasing our ability to do so. And given that announcements of astonishing new breakthroughs are becoming so common that they seem almost routine (ten years ago, who would have predicted that the first implantable human cloned embryo would have already been created?), it is dangerous to take anything for granted. The time to stop the transhumanist agenda isn't when it is being put into effect, but now, before it is able to get off the ground. And the surest way to accomplish this task is to place all human cloning under a ban.

The New Eugenics

Even if science cannot yet—and may never—grant the transhumanists and would-be genetic enhancers *carte blanche* to do what they want, their ideology has already damaged our perception of children and what it means to become a parent. Childbearing and rearing seem increasingly to be viewed as being primarily about *us*—satisfying our desires, working toward our own fulfillment through our children's lives. And now, we are deemed by the bioethics elite to have a "procreator's right" to design our children to achieve these goals.

Even before the age of genetic engineering, we have embraced a "decentralized, do-it-yourself eugenics." Genetic counseling, followed by eugenic abortion to eradicate children with Down's syndrome and other disabilities, has become a routine part of prenatal care. Some now hope this practice of eugenic "screening-out" will expand beyond the "therapeutic" to using embryo screening and selection to increase the likelihood of getting a "better" child.[46]

Oliver Morton enthusiastically embraced such a prospect in his 1998 article "Overcoming Yuk":

> A couple goes to a clinic and provides some sperm and some eggs. The clinic turns them into embryos and analyzes the different mixtures of the parents' genes each embryo carries. The parents are given the embryos' genetic profiles and advice on how the genes relate to various traits, both physical and mental, in various different conditions.... The parents choose the profile they like, or whatever criteria appeal to them; the chosen embryo is grown a bit further in the test tube, a few cells are snipped out to provide tissue for repairs in later life, and then the pregnancy gets under way. No engineering; just choice.[47]

But many worry about such breezy acceptance of this new, more effective eugenics. Ruth Hubbard and Stuart Newman, writing in the March 2002 edition of Z Magazine, warn that we are creating "a world with genetic haves and have-nots." The first step, what they call "choice eugenics," has already been taken. But now, a new, and for them more disturbing, "Yuppie-eugenics" is evolving:

> What was once a preventative choice has become a pro-active entitlement, exacerbated by the sense prevailing among current elites that one has the right to control all aspects of one's life and shape them by buying and periodically upgrading the best that technology has to offer, be it a computer, a car, or a child. Because this trend enjoys broadly based, mainstream sanction in the United States, what may begin as elite yuppie-ism is poised to become more widely disseminated as the technologies become cheaper and their use becomes more routine.[48]

Are such concerns overblown? I think not. The basic principle that children can be created, at least in some situations, as "a means to an end" is already accepted. In one instance, British parents who were not otherwise planning to have additional children received permission from a government ethics panel to undergo IVF to create a baby whose umbilical-cord blood could be used to treat an ill sibling.[49] Supporters say this was a win-win situation since the new child was loved and valued. That is true. But what if parents undertook embryo selection with no intention of welcoming the resulting baby into their family? What if, on the basis of tissue typing, they selected an embryo simply to be used as a fetal organ donor—that is, to be gestated but not brought to birth? Or what if, after recovery of umbilical-cord blood, the baby was unwanted and turned over to an adoption agency? Would we still have a positive view of creating children as a means to a medical end?

Conception by design has already transcended therapeutic purposes. Rich couples pay to obtain eggs or sperm that they hope will provide the best-looking and most intelligent progeny money can buy. This commercialization of gametes has actually driven the price of human eggs through the roof, particularly those taken from "supermodels" and young, beautiful women who attend elite colleges. In 1999, for instance, "eugenically superior" eggs commanded up to $50,000 in an Internet auction.[50] There is even a website where you can buy sperm from intelligent hunks, and there has long been a sperm bank where would-be procreators can buy the seed of Nobel Prize-winners.[51]

Lest we become complacent that the relatively crude selection methods already being used are not the same as outright genetic engineering, we should not ignore the general trajectory. Pandora's box is opening, with human cloning on the verge of becoming the first monster to escape. Transhumanist Gregory Stock crows in *Redesigning Humans* that once they are unleashed, "these technologies will be virtually impossible to control."[52]

Even now, the *ideas* that could push us toward accepting transhumanist premises are spreading like a stain. Demonstrating the dual uses to which some technologies could be put—i.e., therapeutic or enhancing—"Schwarzenegger mice" have already been created by genetically altering rodents to halt the depletion of muscles during the aging process.[53] With astonishing speed, the eugenic core of transhumanism and related philosophies has regained acceptability in elite, influential academic milieus in the life sciences, moral philosophy and the law.[54]

In the decades ahead, if unrestrained biotech research leads us to the brink of fabricating "designer babies," the values espoused by today's new eugenicists will be implemented in the flesh of tomorrow's children, increasingly distorting family life. In a very telling section near the end of *Who's Afraid of Human Cloning?* Gregory Pence foresees just such an eventuality: children will in the future be selected as we now choose pets:

> When it comes to non-human animals we think nothing of trying to match the breed to the needs of the owner . . . [M]any people love their retrievers and their sunny dispositions around children and adults. Could people be chosen the same way? Would it be so terrible to allow parents to at least aim for a certain type, in the same way that great breeders . . . try to match a breed of dog to the needs of a family?[55]

Opponents of these agendas worry that the increased power of biotechnology, like an updated computer program, would grant the new eugenics far greater destructive power than the original version. But neo-eugenicists disagree: the old eugenics led to despotism primarily because it was governmentally enforced, not because eugenics theory is wrong *per se*; they believe that a *laissez faire* system of genetic "choice" would liberate us, allowing our desires to take us where the collectivity wants to go. Others tout an egalitarian eugenics that would enlist the government and supply taxpayer funding to ensure that all parents have an equal chance to participate in the coming genetic arms race.[56] In *The Lives to Come,* Philip Kitcher announces that the new techniques make "some form of eugenics . . . inescapable." To prevent repeating past mistakes, he recommends shaping the new eugenics into a "utopian" form in which all differences are properly valued and where "citizens would be educated but not coerced" to make good choices about the children they bring into the world.[57]

In *Children of Choice,* bioethicist John A. Robertson, a law professor at the University of Texas, asserts that women, along with an absolute right to terminate their pregnancies, have an absolute right to access whatever "noncoital technology" they require to bear children. This right is so fundamental that it includes a right to meddle with our kids' genetic makeup—a procedure he calls "quality control of offspring."[58]

Robertson, quick to anticipate possible objections, asks whether this might dehumanize our perceptions of children and transform them from flesh-of-our-flesh and blood-of-our-blood into products chosen like goods in a shop window. Yes, that could happen, he concedes. But so what? "Although [embryo] selection techniques will permit some defective 'products' to be repaired before birth, most affected fetuses will be discarded, based on judgments of fitness, worth, or parental convenience."[59]

Robertson views the decision to become a parent through a cool, utilitarian prism, reducing this most profound human experience to the equivalent of achieving a rewarding career goal or pursuing an interesting avocation. Parents are free to improve their progeny as they wish—a right he grotesquely dubs "the fabricator's procreative liberty."[60] Fabricators may "take actions to assure [*sic*] that their offspring have characteristics that make procreation desirable or meaningful for [the fabricators]. On this theory, both negative and positive means of selection would . . . be protected."[61] In other words, parents should have the right to select attributes from a menu of options, personalizing their children into pleasing, wish-fulfilling packages like so many anniversary gifts. Of course, they also have the right to abort an unborn child should prenatal testing determine that he or she is unlikely to fulfill their desires. (Robertson is silent about whether this would extend to active infanticide for disappointed parents—a remedy that Princeton bioethicist Peter Singer, for one, has embraced.)[62]

Robertson was an early voice in favor of these parental rights. In recent years, as the human genome was successfully mapped, other voices have joined the chorus. Their solipsistic mantra permeates a 2003 essay by feminist bioethicist Jean E. Chambers published in the *Cambridge Quarterly of Health Care Ethics,* a prominent professional bioethics journal. Chambers extrapolates from a woman's legal right to abortion a cognate liberty to determine genetically the children she decides to bear. Pointing out the obvious—that a woman may not legally be forced to have embryos implanted after their creation *in vitro* (or, one assumes, via SCNT)—Chambers infers that "the candidate embryos are effectively at the woman's disposal."[63]

What has love got to do with it? Apparently, not much, because Chambers describes the decision to become a mother in the lexicon of purchase, investment and return:

> It is an important decision whether to bestow one's parental investment on any particular embryo. A prospective mother bestowing nine months of physical nurturance and many years of physical, emotional, and social care on another being is a significant investment of herself, her time, and her money. It is a far more important investment than buying a car, and we think car buyers should thoroughly research the differences among

the various models before deciding which one to buy. Of course, this is not to say that a rational woman is shopping for a baby, but both the woman and the car buyer are making decisions about important personal investments, and we should support their ability to use all available relevant information to make the best decision. The popular label "designer baby" trivializes what is at stake in such decisions.[64]

Chambers's transactional view of the decision to become a mother exposes the profoundly depersonalizing forces that have been unleashed against the family in our society, and which are rapidly co-opting biotechnology. Childbearing, in this jaundiced mentality, is merely another important "investment" for which the prospective mother calculatingly shops among her *in vitro* embryos, picking and choosing in order to determine which is the most likely to provide *her* with the most satisfactory return. As for fathers, they are so unimportant that they barely warrant a mention.

But Chambers doesn't go so far as to be a transhumanist. She resists the idea, for example, of adding firefly genes to an embryo to produce translucent eyes; her argument here is that the future child, by being socially ostracized, might well be harmed by such exotic manipulations. And she worries, justifiably, that the sheer complexity of biochemistry defeats safe and accurate prognostications.

But this caution isn't founded in a principled moral belief that genetic enhancement is wrong. Indeed, should it ever become safe to alter a prospective child's genetic makeup, there would seem to be no reason, given Chambers's values, why a mother couldn't have the local biotech outfit fabricate a posthuman child for her—if that's what she wants. Chambers admits that she dreams "of altering the human genome to reduce men's sexual proprietariness, jealousy, and aggression toward women."[65] No wonder George Annas, chair of the Health Law Department of Boston University School of Public Health, worries that human cloning will mark a "turning point for the human species" by leading to the profound risk that our progeny "will come to be treated more like products or commodities than like the unique, priceless children that they are."[66]

Assuming (for the sake of argument) that the government *would* keep its hands off the new eugenics enterprise, the theory that "choice" would prevent oppressive social forces from developing ignores a crucial truth about human culture: Peer pressure and social coercion often have far greater power to control our behavior than do government policies and the law. Consider for a moment how so many parents let the state of their egos hang on the successes or failures of their children: the awful competitiveness exhibited by some soccer moms and sports-obsessed dads, the neurotic behavior of "stage mothers," and the lengths to which some parents go to ensure that their children are enrolled in the best schools; then imagine the competition that would develop to produce the "best" genetically enhanced babies.

Like some surrealistic game of keeping up with the Joneses, the race to breed ever more intelligent, beautiful and talented children would grow progressively more extreme. And as Eric Cohen, executive editor of the biotech journal *New Atlantis,* points out, "The more biologically improved we become, the less willing we may be to accept imperfection—or the imperfect." In the end, we may become the ultimate victims of our own hubris: "The more we come to believe that life can be fixed, mastered, and ordered to our liking, the less prepared we may be for the disorder and disaster inherent in our mortal condition."[67]

If genetic redesign is to come, it is still decades away. Yet the values that would transform children into so many products have already germinated and broken through the surface into our culture. As reported by Bill McKibben in *Enough: Staying Human in an Engineered Age,* two performance artists opened a fake genetic-engineering boutique—they called it "Gene Genies Worldwide"—to "get people to think more deeply" about the potential for human enhancement. The artists, Tran T. Kin-Trank and Karl Mihail,

> outfitted their store with Petri dishes and models of the [DNA] double helix, and printed up brochures highlighting traits with genetic links: creativity, extroversion, thrill seeking, criminality. When they opened their doors, they found people ready to shell out for designer families (one man insisted he wanted the survival ability of a cockroach). The "store" was meant to be ironic, but the irony was lost on a culture so deeply consumer that this kind of manipulation seems like the obvious next step. "Generally, people refused to believe this story was an art project," says Tran. And why not? The next store in the mall could easily have been a piercing parlor or a Botox salon. We're ready. And no one's even begun to advertise yet.[68]

Should such attitudes become widely shared—not an unlikely prospect, considering how successful the old eugenicists were in popularizing their doctrine—imagine the pressures brought to bear on parents to prevent "inferior" babies from entering the world. And imagine the obstacles those deemed genetically deficient might have to face: social ostracism, lost educational opportunities, difficulty gaining meaningful employment, refusal of health insurance, perhaps even denial of the chance to fall in love, marry and have children.

As the new eugenics accelerates, it won't just be the "genetically inferior" who might not make it through gestation, as happens today. With more sophisticated prenatal testing, the same fate could await those babies, otherwise normal, found to have a genetic predisposition toward certain adult diseases, such as breast or colon cancer. And then, as our knowledge about the interaction of genes increases, unborn life thought to possess certain "undesirable" genetic predispositions—perhaps not even health-related—might also be rejected. These unfortunates might include those diagnosed as having tendencies toward obesity, homely features, alcoholism or other addictive behaviors, homosexuality, poor athletic ability, low intelligence, criminal or other antisocial behavior— the eugenic list could go on for pages.

Eugenics is evil because its self-evident truth holds that all men are *not* created equal. Such thinking devalues the lives of disfavored individuals, leading with all the force of gravity to a fundamentally unjust society. Nor would the new eugenics "just" be about eradicating disease and disabilities. Inevitably it would lead to seeking human perfection through genetic redesign. Had such a world existed in years past, Abraham Lincoln's parents might have decided not to have an ungainly child with a predisposition to depression and the hereditary disorder known as Marfan's syndrome.

In 2002, an international conference of scholars and scientists gathered from across Europe to ponder the matter and declared that there is no "right to be born with a human genome that has not been modified by artificial means." They were in favor of developing germ line genetic engineering for the "promotion of human health," "the promotion of social life," and "the promotion of meaning and a meaningful life."[69] Many of the world's most famous scientists and philosophers are enthusiastic about the eugenic possibilities. Nobel laureate James D. Watson, for one, told the *Times* of London that he considers stupidity a "disease" and that he would "get rid of the lower 10 percent" who now "have difficulty in elementary school" by enhancing them for intelligence.[70] And on another occasion he said, "Perhaps adding genes will turn slow learners into whiz kids, or prevent smokers from ever developing lung cancer, or make people HIV-resistant."[71]

On the surface, Bill McKibben admits, such a view represents "a deeply attractive picture." But he warns that "once the [enhancement] game is under way ... there won't be moral decisions, only strategic ones."[72] The desire to obtain a competitive edge for one's progeny by adding a gene here and altering a gene there would be essentially futile. Just as today's top-of-the line computer is quickly outdated, the enhanced baby would, within the few short years it would take to grow into childhood, become genetically inferior to those born later and given even greater enhancements. Tomorrow's impressive 20-point IQ bonus would pale against next year's 40-point increase. Thus, rather than increasing a child's chances to excel in life, the result of a genetics arms race would actually be to set up future failure; older models would find it increasingly difficult to compete against the new and improved humans continually entering the competitive marketplaces of school, college, career, athletics and the arts.

So in the end, what would be achieved? What real triumph would there be in winning a marathon if the success resulted from greater lung capacity and muscle endurance derived through gene enhancement rather than from personal sacrifice and discipline? Why cheer your favorite baseball team when the league winner would be determined by better hitting through chemistry? And what would become of parental pride in their children once parents became "programmers" and their children mere "products"? Parents would be able to take "precisely as much pride in [their child's] achievement as ... in the achievements of [their] dishwashing detergent. It was designed to produce streak-free glassware, and she was designed to be sweet-tempered, social and smart."[73]

Designing instead of begetting children could also deprive them of the chance to decide their own futures. Today's child might be forced by parental ambitions to pursue disliked endeavors for a time. But eventually they can free themselves—abandoning the piano lessons, putting away the baseball glove, dropping the premed major to become a bohemian artist. But how would children rebel against their gene enhancements? In a sense, our children would never grow up and become independent beings. Human freedom would be eroded. "Parents would be inserting genes that produced proteins that would make their child behave in certain ways throughout his life," McKibben writes.[74]

> Such children will, in effect, be assigned a goal by their programmers: "intelligence," "even temper," "athleticism." ... Now, two possibilities arise. Perhaps the programming doesn't work very well, and your kid spells poorly, or turns moody, or can't hit the inside fastball. In the present world, you just tell yourself that that's who he is. But in the coming world, he'll be, in essence, a defective product. Do you still love him unconditionally? Why? If your new Jetta got thirty miles to the gallon instead of the forty it was designed to get, you'd take it back. If necessary, you'd sue. You'd call it a lemon.[75]

Choice?

The assumption that the reanimated eugenics monster will be controllable is naïve and foolish—similar to the blithe assumption of the fictional characters in Michael Crichton's novel *Jurassic Park* that they could contain cloned dinosaurs behind amusement park fences. Once the odious notion is accepted that some people are fundamentally better than others, and that our progeny can be genetically manipulated into so many versions of good, better or best, the state and other powerful institutions will jump on the bandwagon and take it over—just as they did the old eugenics, which began as a private educational campaign to persuade the eugenically fit to have large families.

Joseph Fletcher—whose leadership in establishing the bioethics movement and his promotion of "situational ethics" made him, in my view, one of the most influential twentieth-century American thinkers—promoted this noxious theory more than two decades ago. Fletcher urged that we apply our emerging genetic knowledge to "protect society and families from the burdens of caring for diseased individuals, and from the misery of their simply 'existing' (not living) in a nearby 'warehouse' or public institution."[76]

Fletcher believed, like the bioengineers of *Brave New World*, that enhancement genetic engineering should be celebrated, and in cases of health or disability, genetic cleansing should be socially mandated.[77] He also promoted the forced abortion of genetically defective children, if necessary: "It could be right either voluntarily or coercively to limit pro-creation by prevention either before or after conception—if and when spec-ified genetic diseases or defects are predictable or at risk."[78] He even promoted the *branding* of carriers of undesirable genetic traits, an idea first advanced by Nobel laureate Linus Pauling when he advocated that genetic "carriers should wear a small tattoo on their foreheads as Indians wear caste marks."[79]

Already, even before the science has been perfected, we are begin-ning to see the emergence of the kind of invidious discrimination that is inherent in eugenics thinking. "Like eugenics," warns Edwin Black in *War Against the Weak,* "newgenics [his term for the new eugenics] would begin by establishing genetic identity, which is already becom-ing a factor in society, much like ethnic identity and credit identity. DNA identity databanks are rapidly proliferating."[80]

Currently, such information is used in law enforcement, which, at least from my perspective, is unobjectionable. It's worth noting, how-ever, that a major article in *Nature* advocated universal DNA tests at birth for potential future use in solving crimes, raising the possibility that such information could also be used to weed out the inferior.[81] Along these lines, Robert Edwards, the father of *in vitro* fertilization, favors prenatal screening to decide eugenically who gets to be born. Calling it a "sin" for parents to give birth to disabled children, Edwards told an international fertility conference, "We are entering a world where we have to consider the quality of our children," a world in which biotechnology will become "a tool of social engineering."[82]

Other plans for genetic information-collecting are already being hatched. As Black points out, "Eventually, genetic databases will go far beyond the identification of mere individuals. The science will create family genetic profiles for use in litigation, health, and employment that may function as credit bureaus do today. The day is coming when such family information will routinely be sought in conjunction with employ-ment, insurance, and credit granting."[83]

Powerful institutions are taking action based on genetic testing. For example, a few years ago an HMO informed a pregnant woman that it would pay for an abortion when her unborn child tested positive for cystic fibrosis, but would not cover the infant under the family's med-ical policy if she chose to carry to term.[84]

This trend will only worsen as the technology advances. Indeed, there are already calls for doctors to refuse to bring babies with disabilities into the world. Julian Savulescu of the Royal Children's Hospital, Melbourne, writing in the *Journal of Medical Ethics,* urges physicians to refuse assistance to couples who want to have a baby if the child would be likely to have a subpar life. And toward the end of improving the human gene pool, focusing primarily on increasing intelligence, he believes people should contemplate using reproductive technologies to have genetically unrelated children if having a biologically related child would result in an inferior product.[85] It would be better "to bring up Einstein's clone" than a genetically related "normal child" under such circumstances. "Is this eugenics?" he asks rhetorically. "Yes, it is. But not all eugenics practices are objectionable."[86]

Displaying all the intensity to improve the human race as his eugenics forebears, Savulescu asserts that genetics and human cloning would allow prospective parents and their doctors to

> have more information than ever before on the disease susceptibilities, talents, and personality traits of the future individual. Such information will allow us to make fine-grained distinctions about the value of the individuals which would result. We may be able to say that this embryo is likely to have a better life than that. If medicine has an absolute commitment to maximization, doctors should only offer to bring into existence those individuals who are expected (based on the information available) to have the best lives.[87]

Many of the most impassioned promoters of biotechnology have always viewed the emerging research through the prism of a new eugenics that would have sharper teeth than the original version and thus be more likely to succeed. Joseph Fletcher unabashedly promoted governmental and societal action to impose eugenics on the weak. "Testes and ovaries are in fact social by nature," he wrote, "and it would appear ethically that they should be controlled in the social interest. It makes good sense, then, that when ethically weighed, the right to reproduce is actually only a privilege."[88]

Coming to the sinister point, Fletcher summed up:

> The conclusion here is that 1) having genetic information, we ought to set a minimum quality standard of human health and potential for selective reproduction; 2) we ought not to bring children into the world, if

they fall below the minimum standard; and, 3) this minimum standard *ought to be backed up by law.*[89]

Fletcher, like many eugenicists, possessed shockingly harsh social-Darwinist impulses. He accused modern medicine of "polluting our common gene pool" by interfering with natural selection, and urged that technology be applied to "increase the quality of the babies we make."[90] His goal: to transform humans into *Homo autofabricus* via genetic engineering, enabling us to "turn more and more from creatures to creators."[91]

Joseph Fletcher is dead, but his philosophical progeny live on. Gregory Pence tells readers in the last sentence of *Who's Afraid of Human Cloning?*: "Call me Joe Fletcher's clone."[92]

Like many in bioethics and biotechnology, Pence sees reproduction as an almost unlimited "right," which includes the right to use any technology required for a person—it need not be a couple—to accomplish that end; human cloning ("asexual reproduction") will be one of these technologies. Pence's "strongest argument" for permitting cloning to produce children (CPC) is that parents might be able to confer "a wonderful genetic legacy" on their offspring. This would be a tricky business, Pence acknowledges. Sure, there would be mistakes, but what of it? "There are mistakes in choosing schools, in trying to plan conception of children, in estimating one's capacity to be a good parent, and such mistakes don't justify a policy that bans children."[93] He next brings on the hard eugenics—the ultimate objective of many advocates of CPC. Not only should parents have the right to genetically "enhance" their offspring, "they are *obligated* to do so." Why? "It is wrong to choose lives for future people that make them much worse off than they otherwise could have been."[94]

But who is to say which human life is inherently "better" and which inherently "worse"? Take people with developmental disabilities, as just one example. Are their lives really less worthy than those of their brothers and sisters with more intelligence? People I have known with Down's syndrome have been the most kind, loving, sweet, cooperative people in the world, earnest contributors to humanity who make us better for their presence—not only for what they give to us but because of what they induce us to give to them. Would society really be better off without them?

Pence writes that he opposes state coercion in these matters. But one wonders for how long he will do so. If genetic engineering became accepted as a norm—and indeed a matter of urgent necessity—would the new eugenicists be content to let individuals make the decisions?

Even now, some among the new eugenicists are less than fastidious about keeping the government from meddling with the fabricator's choice. "The state of a nation's gene pool should be subject to government policies rather than left to the whim of individuals," Dan Wikler, professor of medical ethics at the University of Wisconsin, urged in a presentation before a World Health Organization symposium. The completion of the Human Genome Project would make it possible, he said, to promote some genetic qualities such as intelligence and lower the incidence of others: "The state acts on behalf of future generations. [Government action] may be conceivably required by justice itself."[95]

Wikler and his fellow bioethicist co-authors, Allen Buchanan, Dan W. Brock and Normal Daniels, make an even more explicit call for a new eugenics in *From Chance to Choice: Genetics and Justice*. After promoting a "broad and comprehensive freedom in reproductive choice," they propose that the state should have a role in enforcing a positive eugenics by fostering "a climate of opinion" in which people make proper eugenic choices "that are likely to result in the transmission of the most desirable genes."[96]

But for all the abundant talk of "choice," and of tolerating "the occasional parental choice that might result in a net burden on society," the authors of *From Chance to Choice* never take coercive negative eugenics off the table. If faced with a "crisis of degeneration," the government could legitimately pursue policies dishearteningly reminiscent of the bad old eugenics of the last century:

> The state, in our view, does have a legitimate role as guardian of the genetic well-being of future generations. Though it is not currently popular or common to say so, we find the prima facie case for genetic stewardship as persuasive as that for a state role in conserving nonrenewable resources, in ensuring that savings rather than deficits are our descendants' financial inheritance, in engaging in basic research in science and medicine with very long-term payoffs, and in affirming that our waste products do not have the potential to become toxic in the centuries to come.

If concern for our genetic future were expressed in terms of a determination to rid the land of radioactive wastes that might cause harmful mutations, few would take exception. Such an initiative would be regarded as an environmental safeguard, which it is, but it also involves a eugenic role for the state, if needed, fit into the standard categories of legitimate areas of concern for government.[97]

And so the pernicious pattern of eugenics, like looping videotape, has already begun to replay. In the end it could result in the destruction of our vision of universal, inherent human equality. As Erik Baard wrote in the *Village Voice,* the logical outcome of the "new" eugenics is inherent inequality:

> "Men," or even "human beings" won't be adequate labels anymore. Life will have been radically redefined, along with the fundamental events of birth and death that bracket it. Equality will be moot, and enforcing it could reasonably be seen as unjust to beings with categorically different or greater abilities. Blake's words ring here: "One Law for the Lion and the Ox is oppression."[98]

Baard hopes that the new superbeings will allow inferior, unenhanced people to have our rights protected by "grandfathering" us "into their society: "Mostly original substrate humans would be free to live and love as before, to the best of their limited abilities."[99]

How big of them. But in a culture where the ideal of universal human equality no longer holds sway, where enhancing genetic engineering is deemed a moral obligation on all parents, there is little reason to believe that any "inferior" life would be considered sacrosanct, or that we would retain any rights at all.

The would-be directors of human evolution (forgetting that we are the species that built the "unsinkable" *Titanic*) assert against all available evidence that we have the wisdom and omniscient vision to decide which of us is "better" and which "worse," which human traits should be enhanced and which eradicated. But I think that Paul Ramsey, the theologian and early bioethicist who warned against walking this road, had the far better argument. "Mankind has not evidenced much wisdom in the control and redirection of his [material] environment," he noted. This being so, should we expect man to do a better job of "reconstructing himself"? Rather than improve the human condition, we would be more likely to make things worse. "Boundless freedom and self determination" would, Ramsey predicted, "become boundless destruction."[100]

At the very least, the new eugenics would distort the human condition almost beyond recognition. Our self-perception as beings of equal moral worth, our attitudes toward others, the nature of the family and how family members interact, the overarching purposes of our societies—all would be radically changed. In the transhumanist's worship of radical autonomy and scientism's insistence upon the domination of science without limits, poisonous seeds have been planted that could become the hemlock that kills off our hard-won beliefs in the value of human diversity and universal moral equality.

The deconstruction of the philosophy necessary to true human freedom would permit the emergence of a new society, thoroughly imbued with biocracy, and lead directly to the tyranny against which Huxley warned in *Brave New World.*

CHAPTER 7

Food for Thought

. . . .

THE POLITICAL DRIVE TO explicitly legalize the cloning of human life is fast gathering steam. In the name of medical science, the spokespeople for Big Biotech have already cast aside earlier assurances that they only wanted access to leftover IVF embryos; they now openly and energetically agitate for the right to clone human life for use in treating illness, making transplantable organs, studying disease models and understanding embryonic development. And there is a massive effort under way to erase the distinction between ESC research and SCNT, and to recast them as the same thing. Likewise, there is a tactical move afoot to call what is truly reproductive cloning by the vague term "cloning," a sleight-of-hand that allows the propagandists to pretend that they oppose SCNT, when in fact they seek to legalize it.

California and New Jersey have already created an explicit legal right to manufacture cloned human embryos. California's law requires the embryo to be destroyed by the fourteenth day, a line that is no more likely to be hard and fast than the earlier promise to restrict ESC research to leftover IVF embryos. More radically, as we have discussed in some detail, New Jersey now permits human cloning, implantation of cloned embryos, and gestation into the fetal stage and through the ninth month of pregnancy. With this license comes the prospect of mass-producing human life to be an object for laboratory research or perhaps even fetal organ farming—with the accompanying evil of exposing the world's poorest women to exploitive harvesting of their egg cells. Yet proponents disingenuously insist that these ominous laws merely protect and promote embryonic stem cell research.

As if these events weren't enough cause for alarm, much of the cloning research already being conducted today is precisely of the kind that would allow future biotechnologists to create cloned babies to order, with the potential of reducing reproduction to off-the-shelf replication. Cloning would then emerge as just another reproductive technology to allow infertile or homosexual couples to have biologically related children, or to permit parents to choose the genetic attributes of their children. This technology would quickly distort the relationships between parents and children, making the latter a combination of toy, experiment and wish-fulfillment fantasy. And it would give rise to a new eugenics, gratifying the transhumanists who eagerly beat the drum for a posthuman future.

Still, despite the ominous portents, we must not allow our disquiet to cause us to turn away from all aspects of biotechnology—far from it. Most of the research now being conducted is not harmful to our essential moral values or to the integrity of bedrock human relationships. And most biotechnology does not undermine the sanctity and essential equality of human life or the importance of human dignity—by which we mean the intrinsic worthiness of embodied human life, the view that we are all worthy of equal and ultimate respect, simply and merely because we are human.[1] Sanely pursued, biotech will give us a profound new understanding of human biology, as well as a potential cornucopia of new medical treatments and improvements in the natural world, all of it consistent with a reverence for human life.

So we face a delicate task. We need to resist the hegemony of scientism and the seduction of an immoral science conducted for profit—but without becoming antiscience or thwarters of progress. We need to facilitate the pursuit of knowledge and scientific advance, and we need to promote the economic success of a morally sound biotechnology industry, without surrendering to the "soft dehumanizations of well-meaning but hubristic biotechnical 're-creationism'" that Leon Kass warns against.[2]

What We Should Say Yes To

Too often, the debate over human cloning is characterized in the media as a conflict of values between rationalists versus religionists, pro-life extremists and ecological radicals who allow their beliefs to stand in the way of scientific advance. This is a caricature. There is actually a vast array of developments in the new technologies to which anyone concerned about the threats posed to human dignity by cloning, and the resulting commmodification of human life, can give enthusiastic assent.

Adult stem cell and related research

Biotech optimist Michael Fumento, author of the industry-promoting *BioEvolution,* accurately calls these "stupendous stem cells."[3] Most biotech companies are not engaging in human cloning or ESC research, not necessarily because of moral qualms—although some companies have stayed away from controversial areas precisely because of their disquiet about using human life as an instrumentality—but because ASC and related cellular and body chemical approaches to regenerative medicine seem more likely to provide effective medical treatments to suffering patients, or at least promise to do so in a shorter time.

This commitment is leading to exciting potential breakthroughs. In the last quarter of 2003, for example, I received the following reports:

- Five Parkinson's disease patients who received injections of a natural body chemical known as glial-cell-line derived neurotrophic factor (GDNF) experienced significant improvement in their ability to perform daily activities. Three of the patients regained their sense of taste and smell.[4]

- Adult neural stem (progenitor) cells grafted into the brains of newborn mice resulted in "extensive myelin production," which could lead scientists to a future ASC treatment for multiple sclerosis.[5]

- Stem cells from bone marrow have been found to repair damaged muscle. The researchers involved believe that the results are promising for the future use of ASCs in the treatment of neuromuscular diseases such as muscular dystrophy.[6]

- Scientists in Canada have turned adult skin cells into the building blocks of brain cells—opening the way for their use in new therapies for heretofore incurable diseases. If this technique is perfected, our skin cells could become potent agents of cure.[7]

- Bone marrow stem cells were induced *in vitro* to differentiate into islet cells—i.e., pancreatic cells that produce insulin. The researchers claimed that their findings "show that human bone marrow-derived stem cells may serve as a potential source for cell therapy in the treatment of type 1 diabetes. This means that we may one day be able to use a person's own stem cells to reverse diabetes."[8] Meanwhile, juvenile diabetes was *cured* in mice using human spleen cells. The cells migrated to their pancreases, "prompting the damaged organs to regenerate into healthy, insulin-making organs," curing their diabetes.[9]

- Adult stem cells extracted from a patient's muscles repaired damage to the heart after a heart attack. Such treatment may not even require surgery, as Dutch investigators reported success delivering the cells by a catheter inserted into an artery. Six months after the treatment, an MRI showed "a significant thickening of the heart wall near the injection sites." This was but one of a series of successful experiments using ASCs to treat heart damage reported around the world.*

- Researchers have successfully restored some eye functions by extracting stem cells from human eyes, growing them in culture, and transplanting them into mice. Optimistic researchers hope that the technique could provide a cure for blindness within five years.[10]

- Cells from human fat have proven to be bona fide ASCs that look to be useful in regenerative medicine. Indeed, it appears that 62 percent of human fat cells "could be reprogrammed into turning into at least two other different cell types," according to Duke University researchers.[11]

Adult stem cell and related therapeutic approaches are in numerous current clinical trials or used for the treatment of cancers, autoimmune diseases, anemias, bone and cartilage deformities, corneal scarring, stroke and skin grafts.[12] Genetically modified skin cells, altered to produce nerve growth factor and implanted in the brain of eight Alzheimer's patients in an early human trial, appeared to slow the mental decline by half. "If these effects are borne out in larger, controlled trials," researchers

*Researchers still need to determine whether the treatment could cause arrhythmias. "Muscle-Cell Injections by Catheter Repair Heart," *Journal of American College of Cardiology,* December 17, 2003. Researchers in two mouse experiments failed to "replicate earlier studies that seemed to show they could be coaxed into making new heart muscle." Sabin Russell, "Adult Stem Cell Transplants Fail in 2 Studies," *San Francisco Chronicle,* March 22, 2004.

said, "this could be a significant advance in therapies for Alzheimer's disease."[13] Cells derived from the inside of a tooth were cultured into neural cells that might someday help alleviate Parkinson's.[14] Bone marrow stem cells have rebuilt liver tissue, helping to restore the damaged organs of mice.[15]

The thrust of the research now seems indisputable: while certainly not a sure thing, and with the proviso that much work remains to be done in animal and human studies, ASCs look to be potent sources of efficacious treatments and cures in the years to come.

Establishing a National Umbilical-Cord Blood Stem Cell Bank

Stem cells found in umbilical-cord blood are a very hopeful area of regenerative-medicine research; such cells have already been used to cure sickle cell anemia. In one published study, 36 out of 44 children remained disease-free two years after treatment, and the potential for bad side effects appears lower than in using bone marrow.[16] The fatal nervous-system disease in children known as Hurler's syndrome has also been successfully treated using cord blood stem cells.[17] Umbilical-cord blood stem cells have shown benefit in treating animals with stroke, amyotrophic lateral sclerosis (ALS—Lou Gehrig's disease), and spinal cord injury (although much work remains before such experiments can be tried in humans).[18]

The potential of umbilical-cord blood is universally recognized. But more needs to be done toward storing such blood for ready use in experiments and treatments. That's why senators from both sides of the cloning divide have joined together to introduce the "Cord Blood Stem Cell Act" of 2003 (S. 1717), which would establish a National Cord Blood Stem Cell Bank Network "to prepare, store, and distribute umbilical-cord blood stem cells for the treatment of patients and to support peer-reviewed research." The bill would also provide an initial $15 million in federal funding to carry out the mandate of the law's provisions. Given that it is co-authored by human cloning opponent Sam Brownback (R-KS), and cloning-for-biomedical-research backers Dianne Feinstein (D-CA), Orrin Hatch (R-UT), and Arlen Specter (R-PA), this is one area of biotech in which everyone should be willing to row in the same direction.

Biotechnology That Does Not Threaten Human Dignity

This book has warned against the controversial areas of biotechnological innovation that potentially impact the sanctity/equality of human life, with special emphasis on those areas that would reduce human life to a commodity or encourage the new eugenics. But as Michael Fumento and others have made clear, most biotechnological inquiry does not put human dignity at risk.

Fumento's *BioEvolution* provides an excellent overview of the vast scope and breadth of biotechnology, ranging from the creation of new vaccines and cancer cures to genetically altered, pest-resistant and more nutritious food. The author makes a strong case that bioengineering will greatly assist underdeveloped nations in their struggle to create sustainable agriculture beyond the subsistence level.[19]* And if mammalian cloning leads to genetically engineered cows or goats that produce milk containing human proteins that can be used to make medicines less expensive to produce, or if pigs or sheep can be altered so that their organs can be safely transplanted into humans, I for one am in full agreement.

Biotechnology may also provide us with the means to use gene therapy as a way of treating diseases and genetic anomalies. Gene therapy, unlike genetic engineering, would not affect the genetic makeup of progeny. Rather, it would treat disease by inserting healthy genes in place of those responsible for diseases or disabilities.

Unfortunately, gene therapy has so far not fulfilled researchers' dreams. Early apparent successes treating X-linked severe combined immunodeficiency disease ("bubble baby syndrome") soon turned to dust when human trials were stopped after patients developed leukemia-like conditions as side effects.†

---

*Some of my allies in the struggle against human cloning and eugenic genetic engineering, such as Jeremy Rifkin and Bill McKibben, oppose many of these nonhuman areas of biotechnological research on moral and environmental grounds. I respect these voices, but find myself rooting for the technology because of its great potential for human good without risking human dignity or the unleashing of a new eugenics.

†As this is written, research is reportedly soon to restart.

Even though the potential of gene therapy has not yet been achieved in human studies, there have been notable successes in animal models. For example, mice with chemically induced Parkinson's were treated successfully with gene therapy, restoring their limb movements. (The gene therapy had been delivered using a harmless virus.)[20] Similarly, scientists used gene therapy to assist guinea pigs regenerate a type of inner-ear cell important to the sense of hearing.[21] This important research will continue, very probably creating powerful weapons in the battle against genetically based diseases and disabilities.

Obtain Cloned ESCs Without Destroying Human Life

Today, the only way to get cloned ESCs is from cloned embryos. As far as we know, this feat has been done only once. But on the horizon there may be another way of obtaining them.

William Hurlbut, a consulting professor of biology at Stanford University and member of the President's Council on Bioethics, believes that scientists might be able to manipulate eggs and somatic cells and fuse them into a tissue mass that would develop cells akin to those derived from cloned embryos—without their ever having been part of a developing human life.

These cells would constitute a "clonal artifact," something that Hurlbut—who would legally ban human SCNT—stresses could be made without it ever having been an embryo capable of "organic functioning." Theoretically, these biological artifacts would produce pluripotent stem cells, but since they would have been produced without creating and destroying human life, experimental use of such tissues would not pose any of the ethical dilemmas associated with human cloning or ESC research. Hurlbut optimistically hopes that progress in this area could convince the biotech vanguard to agree to a moratorium or ban on human SCNT.[22] (In preliminary discussions, he has received positive reactions about the project from many prominent scientists.)

While I am pleased that some prominent biotechnologists would support Hurlbut's project, I doubt they would agree to put human cloning on hold, given the ideology-driven zeal of many cloning advocates and their intensely held belief that only they are entitled to make ethical judgements about the boundaries of scientific inquiry. Indeed, I doubt if *any* conceivable substitute for human cloning would persuade the ideologues of scientism willingly to restrict research activities for very long.

Still, even if the biotech community refuses to go along, Dr. Hurlbut's "third way" could have a tremendous impact on the general community, which wants the benefits of an advanced medical science but also has profound reservations about human cloning. As a believer in the rigorous pursuit of science within the context of moral propriety, I support Hurlbut's imaginative approach. If he could demonstrate that his solution is technically feasible and not cloning by another name, there seems every reason to say Yes, and no reason to say No.*

What We Should Say No To

In March of 2004, bioethicists, biotechnologists and many in the media flew into a tizzy when two members of the President's Council on Bioethics were not reappointed at the expiration of their terms. Proponents of human cloning were enraged, claiming that the two, William C. May and Elizabeth Blackburn, had been forced out because they voted in the council's first report, *Human Cloning and Human Dignity,* against placing a legal moratorium on cloning-for-biomedical-research.†

The *Washington Post's* reasonable editorial statement was in the minority: "The Council is unusual both because its membership includes 'right-to-lifers' as well as some of the country's best-known scientists, and because its nuanced, painstaking public reports have tried to accommodate the views of all, albeit not always to everyone's satisfaction." The *ad hominem* attack made by Timothy Noah in *Slate* was more typical, accusing council chairman Leon Kass of intentionally tilting the council to reflect his views and, moreover, calling him a "silly ass" and a "liar" for denying it.[23]

Although it was more vituperative than most, this was but another skirmish in a far deeper and more profound cultural struggle being waged

---

*Dr. Hurlbut's remarks about embryos are interesting. Human embryonic lives deserve our protection because they "contain within themselves the organizing principle of the full human organism. This is not an abstract or potential ... but rather a potency, an engaged and effective potential in process, an activated dynamic of development in the direction of human fullness of being." "Statement of Dr. Hurlbut," *Human Cloning and Human Dignity: The Report of the President's Council on Bioethics* (New York: Public Affairs, 2002), pp. 310–11.

†In fact, May was not forced out. He was retiring at age 76.

to determine which moral views will drive public policy on bioethical issues in the coming century. The outcome of debates such as this may well decide whether science will serve society through the pursuit of knowledge conducted within acceptable ethical bounds, or come to dominate it by establishing a legal right to do almost anything the research community desires. Sometimes, it will be necessary to just say no—particularly to:

Human Cloning

Political commentator Charles Krauthammer was right on target when he argued in a recent article:

> The problem, one could almost say, is not what cloning does to the embryo, but what it does to us. Except that, once cloning has changed us, it will inevitably enable further assaults on human dignity. Creating a human embryo just so it can be used and then destroyed undermines the very foundation of moral prudence that informs the entire enterprise of genetic research: the idea that, while a human embryo may not be a person [a concept I reject], it is not a nothing. Because if it is nothing, then everything is permitted. And if everything is permitted, then there are no fences, no safeguards, no bottom.[24]

It's hard to disagree with Gregory Stock's starry-eyed thesis in *Redesigning Humans* that biological enhancement will challenge "our basic ideas about what it means to be human."[25] And Leon Kass is equally correct when he writes, as if in response, "To turn a man into a cockroach—we don't need Kafka to show us—would be dehumanizing. To try to turn a man into more than a man, might be so as well. We need more than generalized appreciation for nature's gifts. We need a particular regard for the special gift that is our own nature."[26]

If the years I have spent researching these issues have convinced me of anything, it is the urgent need for all societies to forbid human cloning. The first step is to prevent somatic cell nuclear transfer from being performed with human tissues and cells. However, we should not ignore the point made vigorously and often by the bioethicist Dianne Irving—that there are other potential ways to clone mammalian life than SCNT and that banning it may not be enough.[27] We will have to work unflaggingly to ensure that the laws prevent biotechnologists from finding and exploiting loopholes.

Prohibiting human cloning will be beneficial in many ways. First, it will deprive researchers of the funds they require for their costly experiments. Yes, there will be rogue scientists who, in search of international headlines, will pursue their work in underground labs. But they will be pariahs, whose activities will be rightly regarded as a mix of the cranky and the criminal. Private financiers already shy away from cloning research; and illegalizing it would finish the job once and for all by eliminating government financing, now being sought by cloning advocates under the cloak of stem cell research. Above all, with this more than any other public policy, we will check the advance of scientism as an intellectual force in our cultural life and interrupt the march toward Brave New World.

On a positive note, once human cloning is finally placed beyond the pale, it will free up additional financial resources to explore the numerous and promising avenues of research that have such tremendous potential to improve life on this planet in the coming decades.

Germ Line Engineering

We should not allow the genetic alteration of human germ cells (sperm or eggs) or human embryos through the insertion into them of foreign genes or chromosomes—whether human, animal or artificial.

We should also prohibit the intentional fabrication of human/human chimera embryos by methods such as combining two or more embryos into one modified individual.* (This occasionally happens by accident as a result of IVF procedures.) "Ooplasm transfer," a form of chimeric embryo creation by which cellular material from a donor egg is injected into a would-be mother's egg cell and then fertilized, resulting in an embryo with the DNA of three people rather than two, has already been outlawed by the FDA.[28]

Some resist this tough-minded recipe, averring that germ line engineering would be primarily used to eradicate certain disease-causing genetic traits. But we can already prevent genetic diseases from passing down through the generations by voluntarily refraining from having children. Another method, although some consider such actions immoral, is to use IVF to conceive and implant only embryos without genetic dis-

---

*A chimera is an organism consisting of genetically different cells.

ease. Tay-Sachs disease, a genetic condition that afflicts a tiny minority of certain Jewish populations, causing babies to be born mentally impaired, blind, deaf and unable to swallow, and leading to death by age five, has been all but eliminated by these means.[29] Moreover, even if germ line genetic engineering began with a therapeutic rationale—namely, to eradicate disease—as the science progresses its use will inevitably expand to "enhancement." As we have seen, one of the primary goals of transhumanists is to use germ line genetic engineering to enhance their progeny in unnatural ways, such as by adding artificial or animal chromosomes.

The best way to stop eugenic human germ line alterations of the human genome is to prevent the technology from ever being used in human beings. The Council for Responsible Genetics (politically left-leaning) has laid it on the line: "All people have the right to have been conceived, gestated, and born without genetic manipulation."[30]

Patenting Human Life

The biotech business depends on companies being able to protect their inventions through patenting, which gives them the exclusive right to license and develop their technology. Some companies are already claiming the right to patent bioengineered human life. For example, the *National Journal* has reported that Michael West, the head of Advanced Cell Technology, believes "a cloned embryo of less than 14 days, or perhaps one that hasn't developed a brain, is not human but merely cellular life that can be owned and patented."[31]

Since the demise of slavery, the ownership of human life has been strictly forbidden in this country, and we have never in our most malign fantasies contemplated patenting it. Since profit is clearly the motive for such patents, here we have an opportunity to constrain the power of biotechnology within acceptable moral boundaries. We can do this simply by prohibiting companies or individuals from claiming bioengineered human genomes as their "inventions."

This is not a matter for future consideration. The University of Missouri has already received a patent for a technique involved in "mammalian" cloning.[32] And, while the patent does not include the word "human," we are mammals. It is notable that the patent does not exclude human mammals.

The International Center for Technology Assessment also uncovered patent applications before the Patent and Trademark Office (PTO) that, if granted, would either patent human life, or possibly be construed to do so. As disclosed in its publication *Patent Watch*, the following applications are pending:

- "Patent application serial number 09/816,971, assigned to Geron Corp., which claims a 'reconstituted animal embryo' described as having 'its main use in . . . mammalian embryos, particularly ruminant, human, or primate embryos.'"
- "Patent application serial number 09/755,204, assigned to the University of Connecticut, claiming any 'animal' embryo made by a cloning process." Humans, of course, are animals and thus the patent might be construed as including us.
- "Patent application serial number 09/828,876, assigned to the University of Massachusetts and exclusively licensed to Advanced Cell Technology, claiming a mammalian 'fetus obtained according to' a particularly specified cloning process. This patent application specifically contemplates the use of human tissues derived from cloned human embryos, fetuses, and *offspring*, for transplantation purposes."[33]

Patents on human life should be disallowed as against public policy. Patents relating to the creation of "mammals" or "animals" should be construed as excluding humans. After finding a reference to human cells in a patent granted to Edinburgh University, the European Patent Office added restrictions to its terms to prevent the invention from being construed as covering human cloning.[34] Similarly, the United States Congress has quasi-banned the patenting of human life by denying the PTO any funds "to issue patents on claims directed to or encompassing a human organism." Thus, while it is not explicitly against the law to patent human life—although some argue that the question is covered by the 13th Amendment's prohibition on human servitude—the patents will not be issued without formal funding.

Powerful forces in science and industry oppose human patent bans. The Biotechnology Industry Organization (BIO) criticizes the patent funding ban for being "vague" and creating "uncertainty about the PTO's definition of a 'human organism.'" It warns that it could "halt investment and research into developing biotechnological products."[35] More ominously, some in the research community strongly oppose the very idea of limiting patents on the basis of moral principles. Thus, an April 2003 editorial in *Nature Biotechnology* asserted, in reaction to a ruling by the Supreme Court of Canada halting the patenting of human/mouse chimeras, that moral principles have no place in patent law:

> At present, human beings are unique in being the only living creatures that remain off limits to patents. No country's patent system has yet found a way of extricating itself from the philosophical and political morass associated with patent applications that encroach on definitions of humanness. . . . But moral standards are clearly an unsatisfactory benchmark for establishing patentability: morality (like obscenity) is one of those things that arbiters (more specifically, patent examiners) are likely to have a hard time defining. Clearly, better definitions are needed. One potential criterion, for example, could be to reject patent applications on any product that requires the use or inclusion of human embryos over 14-days old (the point at which development of the nervous system and potentially human sentience begins). . . . But legislation clarifying the scope of patents on higher forms of life should steer clear of moral and ethical definitions. We need to stick to rational and scientific benchmarks that can be practically applied by patent agencies.[36]

We must reject such elitism. No scientist, university, or corporation should be able to own any human life. Period.

Some scientists worry that a barrier against patenting human life could prevent the patenting of stem cell production methods, cell lines, transgenic animals capable of making human proteins, and so on.[37] But none of these forms of biotechnological enterprise constitute the making of "human organisms." Thus, such objections are more like a way of liberating Big Biotech and the science establishment from reasonable social regulation and oversight, than expressions of a well-founded professional worry that research might be constrained.

Implanting Embryos for Any Purpose Other Than Giving Birth

In an earlier time, forbidding such an act would have caused a roll of the eyes and a smile. Who would ever implant an embryo in a womb for any other reason than to bring the resulting baby to birth? Who would treat human life so crassly, so inhumanely, as if it were nothing but so much meat?

Unfortunately, we can no longer assume that gestating for research purposes is generally frowned upon. It is clear that some biotechnologists look forward to the time when cloning technology has advanced to the point that they will be capable of implanting cloned or genetically altered embryos in real, animal or artificial wombs for the purposes of study, experimentation or organ harvesting. Indeed, as we've noted, New Jersey has enacted a law that permits these very acts.

Implanting a human embryo for experimental purposes is a line we should never cross. And while we're at it, since biotech advocates seem so averse to exercising self-restraint, the implanting of any human embryo into the body of any member of a nonhuman species or into an artificial womb—unless as part of a medical treatment intended to save a baby threatened by a failing pregnancy—is a step that should be permanently criminalized.[38]

Prevent Human Eggs and Embryos from Becoming Commodities

When Canada banned all human SCNT cloning in March 2004, in an omnibus bill governing reproductive technology, it also prohibited the selling of human eggs and sperm.[39] This was deemed necessary in order to stanch the growing commercialization of human gametes for use in fertility procedures—and the potential health impact on egg donors.[40]

The Canadians weren't necessarily being alarmist. We've seen how in New Jersey, human egg cell prices had soared to $7,500 by October 2002.[41] And since researchers will continue to burn through thousands of eggs cells as they seek to perfect human SCNT—it took Woo Suk Hwang 242 eggs, taken from 16 volunteers, to manufacture 30 cloned blastocysts, leading to just one stem cell line—the demand for them can only rise precipitously.

Genetic Discrimination

With the mapping of the human genome, the increasing use of genetic testing and the proliferation of DNA databases, comes the fear that the more we are able to learn about the individual's biological core, the greater the chances that some of us could be subjected to genetic discrimination. Health or life insurance companies might refuse to insure people with perceived propensities for certain diseases; likewise with employers, who might seek, for instance, to avoid hiring workers likely to come down with cancers. And indeed there have already been instances of apparent discrimination. The *Washington Post* reported that a healthy boy, who had taken prescribed medication to eliminate all risk of a predisposition to a heart disorder, was ruled "genetically ineligible for health insurance coverage."[42]

Politicians have begun to sit up and take notice of the threat of "DNA discrimination."[43] President Bush, and leaders on both the conservative and the liberal sides of the political spectrum, want to prohibit it. Bill Clinton, for example, signed an executive order in 2000 prohibiting the federal government from using genetic information in employment decisions.[44] This is an issue that will require more debate and legislative fixing in the years to come.

What We Need to Talk About More

Biotechnology, like all scientific knowledge, transcends borders. The debates occurring in the United States are worldwide, and with various results. Canada, Australia, Germany, Norway, Taiwan and other countries legally prohibit all human SCNT, while the People's Republic of China, Saudi Arabia, Belgium, the UK and Israel permit cloning-for-biomedical-research. Most nations, including the U.S., have as yet no national policy. If we are to maintain control over these powerful scientific innovations instead of the other way around, effective regulation must apply worldwide.

In the end, these matters will have to be the subjects of binding international agreements. The U.N. General Assembly is expected to vote in the fall of 2004 on whether to prohibit all human SCNT (the "Costa Rica" approach), or only cloning-to-produce-children. If the former passes—more countries than not support it—it will be a good start. But greater efforts will have to be made in the years ahead to ensure that international accords keep pace with the growing breadth of biotech invention and enterprise.

Animal bioengineering is another gigantic problem. What limits should be placed, if any, on the amount of human genetic material that can be introduced into animals? In principle, I do not oppose creating transgenic animals—e.g., with an added human gene—to produce proteins for use in "pharming" for medical products or to create organic models to study human disease. As William Hurlbut, a member of the President's Council on Bioethics, pointed out to me, a chromosome or a gene *per se* is not the locus of human dignity or moral worth.

But there must be a limit to the amount of human DNA that we place into animals' genomes. Since radicals are already urging the creation of novel creatures straight out of H. G. Wells's *The Island of Dr. Moreau*—Gregory Pence believes that we could "improve" higher mammals "to the point where they could communicate better and tell us whether they were thinking"—we should not delay discussions about how far we should go in this direction.[45]

And we should have a national conversation about where we are best advised to invest limited public tax moneys, and in what amounts. Columbia University's Robert Pollack worries that "millions of government dollars are poured into genomic research that is aimed primarily at defining genetic risk for disease—information that is useful but that does not prevent the emergence of disease in the community," dollars that could better be used "to give all children ... free vaccines" to prevent disease.[46]

It is also disturbing, but true, that as we argue about how to use biotechnology to extend life far beyond the three-score-and-ten the Bible says we are allotted, vast swaths of humanity continue to be killed or disabled in childhood by infectious diseases. Sarah Sexton notes in *World Watch*, "While pneumonia, diarrhea, tuberculosis, and malaria account for more than one-fifth of the world's disease burden, they receive less than 1 percent of the funds devoted to health research." There are other urgent health-related areas of research that compete for public funding, such as AIDS, cancer, improving access to health care, and promoting better services for people with disabilities. I take no position on the particulars of these matters, but believe we need to think very carefully about triaging public investments in medical and health research before we invest in more exotic biotech approaches that would primarily benefit the lucky few and the very rich.

The time may come when we should reconsider committing *any* public finances to support ESC research. People who make politically incorrect statements like the above are all too likely to be chastised by Big Biotech and accused of being uncompassionate and uncaring. But if the current pace of promising advances in adult stem cell related areas of research continues to accelerate in the coming years, we may decide that the wisest use of America's public's resources is to put the money into the basket that delivers—the one labeled Nonembryonic. After all, if it's cures we're after, and ASCs can bring them to medical clinics sooner—and without moral misgivings—investing in adult cell therapies is very likely the most responsible way to go.

This is not to say that there are no other potential scientific uses for ESCs, even if they don't become sources of the promised breakthrough medical treatments in the near future. As the President's Council on Bioethics concluded, "They are useful in unraveling the complex molecular pathways governing human differentiation." In addition, embryonic stem cells "could be used to test new drugs and chemical compounds for toxicity," perhaps making it possible to "reduce or eliminate the use of live animals in such testing protocols."[47]

These less direct and immediate pluses could change the moral equation. Currently, the majority of Americans support ESC research because they hope it will provide effective treatments for diseases in the relatively near term. But should this payoff continue to prove elusive, and should the ancillary uses to be derived from the cell lines be less directly beneficial to people with illnesses and disabilities, citizens could decide that the moral freight outweighs the potential scientific gain. At the very least, it would reduce their support for public funding, especially with health care resources being stretched by other priorities, such as the coming to Medicare age of the baby-boom generation.

Conclusion

All of the controversies we have explored in these pages boil down to an essential question: Does individual human life have inherent value simply because it is human? If we answer Yes, as I believe we must if we are to retain our moral equilibrium, we should exercise some badly needed self-restraint and reject human cloning, germ line genetic engineering, the new eugenics and transhumanism as unethical technologies and philosophies that lead to the objectification of human life.

But if we decide to sacrifice the sanctity/equality of human life on the altar of biotechnological power, if we diminish or abandon our belief in the intrinsic value of human life as an objective good, if we hand over judgments about human worth to the subjective opinion of those with political power, then we have started down the slope. The excommunication of the smallest forms of human life from the family of man is but the beginning. When we eschew the principle of intrinsic human worth, we are taking a step toward a world of disposable castes of weak and defenseless humans—born as well as unborn.

With the development of biotechnology, we find ourselves at one of the most important moral crossroads in history. We can pursue biotechnology to treat disease and improve the human condition, while retaining sufficient wisdom to keep ourselves from infringing on the intrinsic value of human life. Or driven by our thirst for knowledge, the lure of riches, our obsession with control and hopes for vastly expanded lifespans, we can foolishly rush down the antihuman path warned against by Aldous Huxley.

These issues are too apocalyptic to be "left to the scientists." Nor can we afford to allow the marketplace to determine what is right and what is wrong. The stakes are too high, the impact on each of us and on society too profound, the implications for our progeny too irrevocable, to remain passive and indifferent to the decisions we face. It is our right and our duty to participate in the historic cultural and democratic debates over biotechnology in which our polity is now immersed. The human future depends on it.

Glossary

· ·

Adult stem cells (ASCs): Undifferentiated cells found in many tissues of the body. They are thought to be potential sources for regenerative medical treatments.

Biotechnology: The use of living organisms or cells for commercial purposes, such as the development of medical treatments or genetically modified crops.

Biotechnology Industry Organization (BIO): The trade and lobbying organization that represents the business and political interests of the biotechnology industry.

Bioethicist: Practitioner in the field of bioethics.

Bioethics: A field of philosophy that seeks to determine right from wrong in medical and biotechnological endeavors. While the beliefs of bioethicists differ, the predominant view in the field does not accept the sanctity/equality of human life.

Blastocyst: An embryo that has reached about one week of development.

Chromosome: A threadlike body in the nucleus of a cell that carries the genes.

Cloning: Generally, the act of creating a copy of a biological entity. As used in this book, the term generally refers to asexual reproduction via somatic cell nuclear transfer (SCNT) of an animal or a human being that results in the creation of a cloned embryo.

Cloning-to-produce-children (CPC): Cloning a human embryo via somatic cell nuclear transfer (SCNT) for the purpose of reproduction.

In this case, the cloned embryo is implanted in a woman's uterus and gestated to birth.

Cloning-for-biomedical-research (CBR): Cloning a human embryo via somatic cell nuclear transfer (SCNT) for the purpose of using it in medical or other research. CBR results in the destruction of the cloned embryo.

DNA: A molecule in the form of a twisted double strand (double helix) that is the chemical substance of genes, the units of hereditary information that occupy fixed positions on the chromosomes. DNA is common to almost all organisms and is self-replicating.

Embryo: Nascent life. In humans, an embryo is the name given to the developing human being from conception through the eighth week.

Embryonic stem cell research (ESC research): Biotechnological research that seeks to learn how to use stem cells derived from embryos in regenerative medical treatments or to obtain scientific information. Deriving stem cells from human embryos is controversial because it destroys the embryo.

Eugenics: The study of improving the human race through directed reproduction. The American eugenics movement sought to accomplish this between 1890 and 1940 through "positive eugenics," which sought to convince eugenically "superior" people to have large families, and a "negative eugenics" that aimed to prevent the "inferior" from reproducing by means of forced sterilizations. Eugenics ideology in Germany helped to spark the Holocaust.

Differentiated cell: A cell that is a specific tissue type; e.g., blood, bone, skin. There are over 200 differentiated cell types in the human body.

Fetal tissue research: Taking cadaver tissue from aborted or miscarried fetuses for use in regenerative medicine.

Genes: A segment of DNA that controls hereditary factors.

Genetic engineering: Adding different genes to an organism to alter it in a desired way. In humans, the purpose of "therapeutic" genetic engineering would be to cure inherited disease or disability. "Enhancing" genetic engineering would be the "improving" of a person—by increasing IQ, for example.

Germ cells: Oocytes (eggs) in women and sperm in men.

New eugenics: A contemporary eugenics philosophy that would improve the human race through applied biotechnology, such as genetic engineering.

Oocyte: An egg cell.

Pluripotent cells: Stem cells thought to be capable of transformation into all types of tissues.

Pre-embryo: A political advocacy term for an early human embryo that has not been implanted in a uterus.

Regenerative medicine: Medical treatments that reduce symptoms or provide cures for illnesses and disabilities caused by degenerated or injured organs or body parts. The treatment is designed to help rebuild the damaged part of the body. Adult and embryonic stem cells are believed to be potentially potent sources of regenerative treatments.

Reproductive cloning: A popular term for engaging in human SCNT for the purpose of reproduction.

Scientism: A worldview that sees science as the absolute and only legitimate means of access to truth, as well as the source of salvation for humanity.

Somatic cells: All cells in the body except germ cells.

Somatic cell nuclear transfer (SCNT): The primary form of mammalian cloning, including in humans. SCNT is accomplished by removing the nucleus from an egg and inserting the nucleus from a somatic cell. The genetically modified egg is stimulated with an electric shock. If the cloning succeeds, a one-celled cloned embryo results and embryonic development proceeds in the same way as in naturally created embryos. SCNT was the cloning method used to create Dolly the sheep.

Stem cells: The common name for undifferentiated cells. Stem cells can be derived from embryos, adult tissues and umbilical-cord blood.

Therapeutic cloning: Performing SCNT for purposes of creating embryos for use in biotechnological research or to derive embryonic stem cells for regenerative medical treatments.

Transgenic: An animal or plant containing genes from a different species.

Transhumanism: A philosophy that seeks to enhance human individuals or their progeny through genetic engineering, robotics, nanotechnology and other futuristic technologies.

Acknowledgments

· ·

This book was several years in the making and could not have been completed without the selfless assistance of so many knowledgeable and selfless people who shared so much of their time and expertise. (For those I may have missed, please forgive me.) Thank you so much to: Tom Abate; Joseph Bottum; Eric Chevlen; Jose Cibelli; Anthony Coles; Marcy Darnovsky; Roland Foster; Diane Gianelli; Susanne and Bob Gray; Kathi Hamlon; Richard Hayes; Robert Hiltner; James J. Hughes; Dianne Irving; William Kristol; Michel Levesque; Rita Marker; Leon Kass; Neil Munro; Suzanne Murray; Claire Nader; Stuart A. Newman; Judy Norsigian; Josephine and Bruno Quintavale; Jeremy Rifkin; Joni Eareckson Tada; Marie Tasy; Dennis J. Turner; Clara and Gregg Vest; Rob Wasinger, and Dan Wikler.

Special thanks to Eric Cohen, Richard Doerflinger, William B. Hurlbut, James Kelly, and David Prentice.

I deeply appreciate the commitment and support of the Discovery Institute to all that I am striving to accomplish. Thank you Bruce Chapman, Jay Richards, Steve Meyer, Jonathan Wells, and my other Discovery friends. The Center for Bioethics and Culture has also been a tremendous support and special asset in the creation of this book. Thank you Jennifer Lahl, Nigel Cameron, and Center participants throughout the country. A tip of the hat to Bradford William Short in appreciation of his support.

Muchas gracias to the people at Encounter Books for their professionalism and enthusiasm. Peter Collier, thanks again for excellence in editing.

Finally, my love and gratitude to my friends and family, who help keep me going and put up with my obsessions, especially my mother, Leona, and Debra J., my wife and total sweetheart.

Notes

· ·

Introduction

[1] The announcement by the Raelians that they had brought the first cloned human baby into existence made headlines around the world. For example, see Nancy Gibbs, "Abducting the Cloning Debate," *Time*, January 13, 2003.

[2] Geron Corp. press release, "Geron Grows Stem Cells without Mice Feeder Cells," October 1, 2001.

[3] For example, see Lee M. Silver, *Remaking Eden: Cloning and Beyond in a Brave New World* (New York: Avon Books, 1997).

[4] BBC News, "GM Goat Spins Web-Based Future," August 21, 2000; Kenneth Chang, "In Experiment, Mammal Cells Produce Silk Like a Spider's," *New York Times*, January 18, 2002.

[5] Scott Foster, "Man-Beast Hybrid beyond Talking Stage," *National Post*, August 22, 2001.

[6] Nicholas Wade, "Stem Cell Mixing May Form a Human-Mouse Hybrid," *New York Times*, November 27, 2002.

[7] For example, see Rick Weiss, "Stem Cell Transplant Works in California Case," *Washington Post*, April 9, 2002.

[8] Agence France-Presse, "DNA Discoverer: Use Genetics to Improve IQ," February 28, 2003.

[9] Leon Kass, interview with author, July 19, 2002.

[10] Neil Munro, interview with author, December 16, 2002.

[11] Stephen M. Barr, "Retelling the Story of Science," *First Things*, March 2003, p. 16.

[12] Roger A. Pielke Jr., "Policy, Politics, and Perspective," *Nature*, vol. 416 (March 28, 2002), p. 368.

13   Gwen Kinkead, "Stem Cell Transplants Offer New Hope in Some Cases of Blindness," *New York Times*, April 15, 2003.

Chapter 1

1   Aldous Huxley, Foreword to *Brave New World* (1948; New York: Harper Perennial, 1998), p. xi.
2   Ibid., p. 12.
3   Marie McInerney, "Australian Researchers Claim to Fertilize Eggs without Sperm," Reuters as published in *The Inquirer,* July 10, 2001.
4   Robin McKie, "Men Redundant? Now We Don't Need Women Either," *The Observer,* February 10, 2002.
5   World Transhumanist Association, "The Transhumanist Declaration," www.transhumanism.org.
6   Glenn Zorpette and Carol Ezzell, "Your Bionic Future," *Scientific American,* September 1999.
7   Janet Rae-Dupree, "Know Your Genes, Know Yourself," *U.S. News and World Report,* May 27, 2002.
8   Gregory Stock, *Redesigning Humans: Our Inevitable Genetic Future* (New York: Houghton Mifflin Company, 2002), pp. 194, 195.
9   Gregory E. Pence, *Who's Afraid of Human Cloning?* (Lanham, Maryland: Rowman & Littlefield Publishers, 1998), p. 114.
10   James J. Hughes, "The Future of Death: Cryonics and the Telos of Liberal Individualism," *Journal of Evolution and Technology,* vol. 6 (July 2001). Stock, *Redesigning Humans,* pp. 182–85.
11   Howard L. Kaye, "Anxiety and Genetic Manipulation: A Sociological View," *Perspectives in Biology and Medicine,* vol. 41, no. 4 (Summer 1998), p. 488.
12   Leon Kass, *Life, Liberty and the Defense of Dignity: The Challenge for Bioethics* (San Francisco: Encounter Books, 2002), p. 33.
13   Leon Kass, "Preventing a Brave New World," *New Republic,* May 21, 2001.
14   Lee M. Silver, *Remaking Eden: Cloning and Beyond in a Brave New World* (New York: Avon Books, 1997), pp. 249–50.
15   Huxley, Foreword to *Brave New World,* p. xvii.
16   Kass, *Life, Liberty and the Defense of Dignity,* p. 5.
17   David A. Prentice, *Stem Cells and Cloning* (San Francisco: Benjamin Cummings, 2003), p. 3.
18   Ibid., p. 4.
19   Ibid., pp. 11–12. National Institute of Dental and Craniofacial Research, "Scientists Discover Unique Source of Postnatal Stem Cells," reporting that a "rich supply of stem cells" have been found to exist in the "tempo-

rary teeth that children begin losing around their sixth birthday," press
release, NIH News, April 21, 2003.

20 Source: *Stem Cells and the Future of Regenerative Medicine,* Report of the
National Academy of Sciences (Washington, D.C.: National Academies
Press, 2001), p. 6.

21 Source: A. E. Lang and A. M. Lozano, "Parkinson's Disease: First of Two
Parts," *New England Journal of Medicine,* vol. 39, no. 15 (1998), pp.
1044–53.

22 *The American Medical Association Encyclopedia of Medicine* (New York:
Random House, 1989), pp. 772–73.

23 Morton Kondracke, *Saving Milly: Love, Politics, and Parkinson's Disease*
(New York: Public Affairs, 2001), p. 85.

24 Shota Kodama et al., "Islet Regeneration during the Reversal of Autoimmune Diabetes in NOD Mice," *Science,* vol. 302 (November 14, 2003),
pp. 1223–27.

25 Ibid., p. 1227.

26 Geron Corporation, "Geron Grows Stem Cells without Mice Feeder
Cells," press release, October 1, 2001. [My emphasis.]

27 "The California Stem Cell Research and Cures Initiative," Section 3,
subsection 5.

28 National Bioethics Advisory Commission, *Ethical Issues in Human Stem
Cell Research,* September 1999, p. ii.

29 Ibid., pp. 3, 7.

30 Ibid., p. 53.

31 Gretchen Vogel, "Can Adult Stem Cells Suffice?" *Science,* June 8, 2001.

32 *Washington Fax,* "Stem Cells Won't Be Used As Therapies but Will Spawn
Them, Pioneer Gearhart Predicts," November 19, 2002. The article
reports on the opinions of John Gearhart, director of research for Johns
Hopkins University's Department of Gynecology and Obstetrics, presented in a seminar at the National Institutes of Health.

33 L. M. Bjorklund et al., "Embryonic Stem Cells Develop into Functional
Dopaminergic Neurons after Transplantation in a Parkinson's Rat
Model," *Proceedings of the National Academy of Sciences,* vol. 19 (Febru-
ary 2002), pp. 2344–49.

34 F. Nishimura et al., "Potential Use of Embryonic Stem Cells for the Treatment of Mouse Parkinsonian Models," *Stem Cells,* vol. 21 (March 2003),
pp. 171–80.

35 S. Wakitani et al., "Embryonic Stem Cells Injected into the Mouse Knee
Joint Form Teratomas and Subsequently Destroy the Joint," *Rheumatol-
ogy,* vol. 42 (2003), pp. 162–65.

36 S. Sipione et al., "Insulin Expressing Cells from Differentiated Embryonic Stem Cells Are Not Beta Cells," *Diabetologia,* published online, February 2004, as reported by Do No Harm, www.stemcellresearch.org/facts/factsheet-04-03-02.htm.

37 Rebecca D. Folkerth and Raymon Durso, "Survival and Proliferation of Nonneural Tissues, with Obstruction of Cerebral Ventricles, in Parkinsonian Patient Treated with Fetal Allografts," *Neurology,* May 1996, pp. 1219–25.

38 See, for example, ABC News, "Self-Mended Heart," March 6, 2003.

39 Nicholas Wade, "Doctors Use Bone Marrow Stem Cells to Repair a Heart," *New York Times,* March 7, 2003.

40 Dwayne Hunter, "Lifesaving Stem Cell Procedure Halted by U.S. Agency," Betterhumans.com, June 18, 2003.

41 Nancy Touchette, "Bone Marrow Stem Cell Trial Approved for Heart Patients," *Genome News Network,* April 16, 2004.

42 Theratechnologies, Inc.; Celmed Biosciences, Inc., "Adult Stem Cells Used to Repair Damage from Parkinson's Disease," press release describing report at the annual meeting of the American Association of Neurological Surgeons, April 8, 2002. Interview by author with Michel F. Levesque, M.D., October 30, 2002.

43 Dennis Turner, interview with author, October 21, 2002. Levesque interview.

44 Levesque told me that the FDA has authorized him to conduct further human trials once certain animal studies have been completed and his laboratory is upgraded.

45 Kodama, "Islet Regeneration in NOD Mice," p. 1227.

46 I take no position about abortion in this book or in my public advocacy. When discussing the differing opinions in the abortion debate, I call each side by the name they prefer using to identify themselves: pro-life and pro-choice.

47 George Q. Daley, "Cloning and Stem Cells—Handicapping the Political and Scientific Debates," *New England Journal of Medicine,* vol. 349 (July 17, 2003), p. 211.

48 For more information on the SCNT method, presented in an easy-to-understand way, see David A. Prentice, *Stem Cells and Cloning* (New York: Benjamin Cummings, 2003), pp. 21–22.

49 *Human Cloning and Human Dignity: The Report of the President's Council on Bioethics* (New York: Public Affairs, 2002).

50 Ibid., p. 56.

51 Ibid., pp. 57–58.

52 Ibid., p. 59. Emphasis within the text.

53 Ibid.

54 Ibid.

55 Huxley, Foreword to *Brave New World,* p. xi.

Chapter 2

1 Albert R. Jonsen, "O Brave New World: Rationality in Reproduction," in *Birth to Death: Science and Bioethics,* ed. David C. Thomasma and Thomasine Kushner (Cambridge, UK: Cambridge University Press, 1996), p. 50.

2 Joseph Fletcher, *The Ethics of Genetic Control: Ending Reproductive Roulette* (Buffalo, New York: Prometheus Books, 1988), pp. 3–4.

3 Ibid., p. 126.

4 Ibid., p. 157.

5 Ibid., p. 2.

6 Ibid., p. 173.

7 Ibid., p. 172.

8 Michael Fumento, *BioEvolution: How Biotechnology Is Changing Our World* (San Francisco: Encounter Books, 2003), p. 183.

9 Richard John Neuhaus, "The Return of Eugenics," *Commentary,* April 1988.

10 Jeffrey Brainard, "A New Kind of Bioethics," *Chronicle of Higher Education,* May 21, 2004.

11 Ellen Goodman, "Making Babies," syndicated column, January 17, 1980.

12 Ibid.

13 The "Dickey Amendment" is a limitation amendment to the annual appropriations bill for the Dept. of Health and Human Services. It must be renewed annually. The version of the law currently in effect is Public Law No: 108–7 (omnibus FY 2003 appropriations act).

14 Executive Order 12975, issued October 3, 1995.

15 For a detailed discussion of the predominant moral values and premises of the bioethics movement, see Wesley J. Smith, *Culture of Death: The Assault on Medical Ethics in America* (San Francisco: Encounter Books, 2000).

16 NBAC, *Ethical Issues in Human Stem Cell Research* (Rockville, Maryland, 1999), p. 4.

17 National Institutes of Health Guidelines for Research Using Human Pluripotent Stem Cells, 65 *Fed. Reg.* 51976–81 (August 25, 2000).

18 "Nobel Laureates' Letter to President Bush," as published in the *Washington Post,* February 22, 2001.

19 United Methodist Church, "GBCS General Secretary Calls on President

Bush to Extend Moratorium on Human Embryo Stem Cell Research," press release, July 17, 2001.

20 Do No Harm Coalition, "140,000 Petitions Sent to White House in Anticipation of Bush Announcement Tonight," press release, August 9, 2001.

21 "Nobel Laureates' Letter," *Washington Post*.

22 "Scientists Create New Stem Cell Lines from Donor Gametes," *Highlights from Fertility and Sterility*, vol. 176, no. 1 (July 2001).

23 Sheryl Gay Stolberg, "Scientists Create Scores of Embryos to Harvest Cells," *New York Times*, July 11, 2001.

24 Aaron Zitner, "Embryos Created for Stem Cell Research," *Los Angeles Times*, July 11, 2001.

25 For example, see Mae-Wan Ho, "Adult Versus Embryonic Stem Cells," Institute of Science in Society, London, July 12, 2002.

26 JoAnn D. Eiman, Snowflakes Director of Communications, interview with author, April 15, 2003, and March 18, 2004.

27 Ibid.

28 Suzanne Smalley, "A New Baby Debate," *Newsweek*, March 24, 2003.

29 Susanne Gray, interview with author, March 20, 2003.

30 Clara Vest, interview with author, March 20, 2003.

31 Smalley, "A New Baby Debate."

32 Rick Weiss, "400,000 Human Embryos Frozen in U.S.," *Washington Post*, May 8, 2003.

33 Smalley, "A New Baby Debate."

34 Leon Kass, interview with author, July 19, 2002.

35 MSNBC, *Hardball with Chris Matthews*, June 20, 2001.

36 Keith Moore and T. V. N. Persaud, *The Developing Human: Clinically Oriented Embryology*, 7th ed. (Philadelphia: W. B. Sanders Company, 2003), p. 2.

37 Ibid., p. 16. (My emphasis.)

38 Ronan O'Rahilly and Fabiola Muller, *Human Embryology and Teratology*, 3rd ed. (New York: Wiley-Liss, 2001), p. 8.

39 Helen Pearson, "Your Destiny, from Day One," *Nature*, July 8, 2002.

40 O'Rahilly and Muller, *Human Embryology*.

41 Ibid., p. 7.

42 Lee M. Silver, *Remaking Eden: Cloning and Beyond in a Brave New World* (New York: Avon Books, 1997), p. 39.

43 O'Rahilly and Muller, *Human Embryology*, p. 12.

44 Ibid., p. 88.

45 Silver, *Remaking Eden*. (Emphasis within the text.)

46 Ibid., p. 41.

47   O'Rahilly and Muller, *Human Embryology*, p. 8.

48   Maureen L. Condic, "Life: Defining the Beginning by the End," *First Things,* May 2003, p. 52.

49   Silver, *Remaking Eden*, p. 22.

50   Ibid., p. 23.

51   Discussions about the concept of speciesism are rife throughout bioethics literature. For example, see Peter Singer, *Animal Liberation,* new revised ed. (1975; New York: Avon Books, 1990).

52   For a more detailed discussion of bioethics, bioethicists, and personhood theory, see Wesley J. Smith, *Culture of Death: The Assault on Medical Ethics in America* (San Francisco: Encounter Books, 2001).

53   For example, see Ronald Dworkin, *Life's Dominion: An Argument about Abortion, Euthanasia, and Individual Freedom* (New York: Vintage Books, 1994). Also, Peter Singer, *Rethinking Life and Death: The Collapse of Our Traditional Ethics* (New York: St. Martin's Press, 1994).

54   Herbert Hendin, *Seduced by Death: Doctors, Patients, and the Dutch Cure* (New York: W. W. Norton & Company), 1997.

55   Tom L. Beauchamp, "Failure of Theories of Personhood," *KIEJ*, p. 320.

56   James W. Walters, *What Is a Person?* (Chicago: University of Illinois Press, 1997), p. 113.

57   R. Hoffenberg et al., "Should Organs from Patients in Permanent Vegetative State Be Used for Transplantation?" *The Lancet,* vol. 350 (November 1, 1997), p. 1321.

58   Robert D. Truog, M.D., and Walter M. Robinson, M.D., "Role of Brain Death and the Dead Donor Rule in the Ethics of Organ Transplantation," *Critical Care Medicine,* vol. 31, no. 9 (2003), pp. 2391–96.

59   Neil Munro and Mark Kukis, "Brave New World?" *National Journal,* May 22, 2004. (The author was also named in the article.)

60   CNN.com, "Christopher Reeve on Politics and Stem Cell Research," July 30, 2001.

61   Suheir Assady et al., "Insulin Production by Human Embryonic Stem Cells," *Diabetes,* vol. 50 (August 2001), pp. 1–7.

62   Jane S. Lebkowski et al., "Human Embryonic Stem Cells: Culture, Differentiation, and Genetic Modification for Regenerative Medicine Applications," *Cancer Journal,* vol. 7, supp. 2 (November/December 2001), pp. S83–93.

63   Testimony of Dr. David A. Prentice, Ph.D., before Colorado State Legislature House Civil Justice & Judiciary Committee, January 29, 2002.

64   "Stem Cells Won't Be Used as Therapies but Will Spawn Them, Pioneer Gearhart Predicts," *Washington Fax,* November 19, 2002.

65   Robert Lanza and Nadia Rosenthal, "The Stem Cell Challenge," *Scientific American,* May 24, 2004, p. 94.

66   For example, the biotech lobbying group Biotechnology Industry Organization (BIO) advocated "therapeutic cloning" as the cure for the problem of "immunological rejection." See BIO website, "The Value of Therapeutic Cloning for Patients."

67   In an ABC interview, Reeve advocated human cloning to obtain stem cells, "so you avoid the immune system rejection, which is a very key factor." *This Week,* June 16, 2002, FDCH Transcripts.

68   David I. Hoffman et al., "Cryopreserved Embryos in the United States and Their Availability for Research," *Fertility and Sterility,* vol. 79, no. 5 (May 2003), pp. 1063–69.

69   Chad A. Cohen et al., "Derivation of Embryonic Stem Cell Lines from Human Blastocysts," *New England Journal of Medicine,* March 25, 2004.

70   For example, see "Stem Cell Transplant Offers Hope to Children with Sickle Cell Disease," The American Society of Hematology, press release, December 8, 2002; "Stem Cell Therapy Stalls Multiple Sclerosis in Some Patients," Associated Press, April 17, 2002; "Stem Cells Help Regenerate Tissue Damaged from Heart Attack," *Science Daily,* May 29, 2001; "Hope for a Cruel Killer," *Sydney Morning Herald,* November 19, 2002, which reports on a paper presented at the International Symposium on Motor Neuron Disease in Melbourne, describing human clinical trials for ALS patients by Italian researchers using the patients' own bone marrow stem cells.

71   Christopher Reeve, testimony before Senate Health, Education, Labor and Pensions Committee, transcript, March 5, 2002.

72   Source: Amyotrophic Lateral Sclerosis Association website: www.alsa.org.

73   N. Ende et al., "Human Umbilical Cord Blood Effect on SOD Mice (amyotrophic lateral sclerosis)," *Life Sciences,* vol. 67 (May 26, 2000), pp. 53–59; R. Chen and N. Ende, "The Potential for the Use of Mononuclear Cells from Human Umbilical Cord Blood in the Treatment of Amyotrophic Lateral Sclerosis in SOD1 Mice," *Journal of Medicine,* vol. 31 (2000), pp. 21–30.

74   Reeve testimony before Senate Subcommittee on Health, transcript, March 5, 2002.

75   "Cloning Fact," June 11, 2002, Americans to Ban Cloning, citing (among others): C. P. Hofstetter et al., "Marrow Stromal Cells Form Guiding Strands in the Injured Spinal Cord and Promote Recovery," *Proceedings of the National Academy of Sciences,* USA 99, 2199–2204, February 19, 2002; M. Sasaki et al., "Transplantation of an Acutely Isolated Bone Marrow Fraction Repairs Demyelinated Adult Rat Spinal Cord Axions,"

Glia, vol. 35 (July 2001), pp. 26–34; S. Shihabuddin et al., "Adult Spinal Cord Stem Cells Generate Neurons after Transplantation in the Adult Dentate Gyrus," *Journal of Neuroscience,* vol. 20, pp. 8727–35; A. Ramon-Cueto et al., "Long-Distance Aconal Regeneration in the Transected Adult Rat Spinal Cord Is Promoted by Olfactory Ensheathing Glial Transplants," *Journal of Neuroscience,* vol. 18, pp. 3803–15.

76 For an example of Kelly's writing on these issues, see James Kelly, "Cloning: Between Hype and Hope," *Manitou Magazine,* Winter 2004.

77 James Kelly, interviews with author, February 20, 2003, and May 24, 2004.

78 Laurance Johnston and Sara Sá, "Within the Realm," *Paraplegia News,* March 2003, where it is reported that in early human trials, "six of seven patients regained some sensation and muscle control within a month of surgery." Nine months post surgery one patient had regained some bladder control and could stand up using leg braces.

79 The White House, "Remarks by the President on Stem Cell Research," August 9, 2001.

Chapter 3

1   Dana Canedy, with Kenneth Chang, "Group Says Human Clone Born to an American," *New York Times,* December 28, 2002,

2   Despite repeated promises to provide proof of the clonings, not a shred of evidence has ever been presented by Boisselier (or her company, Clonaid, or the Raelian cult to which she belongs) verifying that any cloned babies have been born.

3   As an editorial in *Science* stated, "Legitimate scientists submit evidence, sufficiently substantial to withstand rigorous expert review, to be considered for publication in reputable journals." See "Cloning Claim Is Science Fiction, Not Science," *Science,* vol. 299 (January 17, 2003), p. 344.

4   www.clonaid.com.

5   Raja Mishra, "Little behind Clonaid, Files Reveal," *Boston Globe,* April 23, 2003.

6   From the Raelian website: www.rael.org.

7   Ibid.

8   Jose B. Cibelli et al., "Rapid Communication: Somatic Cell Nuclear Transfer in Humans: Pronuclear and Early Embryonic Development," *Journal of Regenerative Medicine,* vol. 2 (November 26, 2001).

9   Gina Kolata, "Company Says It Produced Embryo Clone," *New York Times,* November 26, 2001.

10 Ian Wilmut, "The Importance of Being Dolly," in *The Second Creation: Dolly and the Age of Biological Control,* by Ian Wilmut, Keith Campbell

and Colin Tudge (Cambridge, Massachusetts: Harvard University Press, 2000), p. 5.

[11] Ibid., p. 6.

[12] An August 2001 poll, published by ABC News, showed that a majority of Americans oppose human cloning for any reason. Two questions were asked: "Should it be legal in the U.S. to clone humans?" Yes, 11%. No, 87% (men 16/82, women 6/93). "Clone humans for medical treatments?" Yes, 33%. No, 63% (men 41/56, women 27/70). A May 14, 2003 Gallup poll disclosed that 90% of respondents believed that "cloning humans" is "morally wrong."

[13] Wilmut, Campbell and Tudge, *The Second Creation,* p. 217.

[14] Jonathan Leake, "Gene Defects Emerge in All Animal Clones," *The Times* (London), April 28, 2002.

[15] Jane Wardell, "Dolly the Cloned Sheep Put to Death," Associated Press, February 14, 2003.

[16] Source: Americans to Ban Human Cloning, info@cloninginformation.org, April 15, 2002.

[17] Helen Pearson, "Pig Fatalities Highlight Cloning Dangers," *Nature Science Update,* August 27, 2003.

[18] Peter Mombaerts, "Therapeutic Cloning in the Mouse," in *Proceedings of the National Academy of Sciences: Regenerative Medicine,* papers from the Arthur M. Sackler Colloquium, held October 18–22, 2002, at the Arnold and Mabel Beckman Center of the National Academy of Sciences and Engineering in Irvine, California (Washington, D.C.: National Academies Press, 2003).

[19] Tom Spears, "Cloning Damages Genes, Harms Health, Warns Expert," *Ottawa Citizen,* September 11, 2002, based on a study by Rudolf Jaenisch, cloning expert at Whitehead Institute for Biomedical Research in Boston, reported in *Proceedings of the National Academy of Sciences.*

[20] Sylvia Pagan, "Cloned Monkey Embryos Are a 'Gallery of Horrors,'" *New Scientist,* December 12, 2001.

[21] Editorial, "Goodbye Dolly … and Friends?" *The Lancet,* vol. 361, no. 9359 (March 1, 2003), p. 711.

[22] As quoted in "Goodbye Dolly."

[23] Spears, "Cloning Damages Genes, Harms Health, Warns Expert."

[24] Rick Weiss, "Efforts to Clone Monkeys Illustrate Potential Hazards for Humans," *Washington Post,* May 10, 1999.

[25] Centre for Genetics Education, "Genetic Imprinting," University of Sydney, Australia, "Fact Sheet 14."

[26] Ibid., p. 222.

[27] J. R. Hill et al., "Clinical and Pathological Features of Cloned Transgenic

Calves and Fetuses (13 case studies)," *Theriogenology,* vol. 8 (1999), pp. 1451–65.

28 James Meek, "Dolly's Creator Says No to Human Cloning," *The Guardian* (Manchester), March 29, 2002.

29 For example, see National Academy of Sciences, *Scientific and Medical Aspects of Human Reproductive Cloning* (Washington, D.C.: National Academies Press, 2002); National Bioethics Advisory Commission, *Cloning Human Beings* (Rockville, Maryland, 1997); Ian Wilmut, Keith Campbell and Colin Tudge, *The Second Creation: Dolly and the Age of Biological Control* (Cambridge, Massachusetts: Harvard University Press, 2000).

30 *Human Cloning and Human Dignity: The Report of the President's Council on Bioethics* (New York: Public Affairs, 2002), p. 105. (Emphasis within the text.)

31 Ibid., pp. 102–3.

32 David A. Prentice, interview with author, May 16, 2003.

33 Lorraine Fraser, "In-Vitro Pioneer Backs Cloning for Infertility," *Sunday Telegraph* (London), June 9, 2002.

34 Ronald M. Green, "I, Clone," *Scientific American,* September 1999.

35 John A. Robertson, "Two Models of Human Cloning," *Hofstra Law Review,* vol. 27, no. 3 (Spring 1999), p. 623.

36 Ibid., p. 636.

37 Gregory E. Pence, *Who's Afraid of Human Cloning?* (Lanham, Maryland: Rowman & Littlefield Publishers, 1998), p. 25.

38 Jeremy Rifkin, "Man and Other Animals," *The Guardian* (Manchester), August 16, 2003.

39 Ted Howard and Jeremy Rifkin, *Who Should Play God? The Artificial Creation of Life and What It Means for the Future of the Human Race* (New York: Dell Publishing Company, 1977).

40 Jeremy Rifkin, interviews with author, January 19, 2002, and December 19, 2002.

41 "Assisted Reproduction May Be Linked to Birth Defects Syndrome," *Science Daily,* November 18, 2002, based on research conducted at Johns Hopkins Medical Institutions.

42 Leon Kass, interview with author, October 8, 2002.

43 Rick Weiss, "Free to Be Me: Would-Be Cloners Pushing the Debate," *Washington Post,* May 12, 2002.

44 Randolfe H. Wicker, "Cloning Ted Williams," ESPN, July 22, 2002.

45 *Human Cloning and Human Dignity,* pp. 114–15.

46 This statement should be qualified, however. See Chapter 1, pg. 16.

47 *Human Cloning and Human Dignity,* p. 117.

[48] Ibid., p. 118.

[49] Ibid., pp. 119–20.

[50] Carolyn Abraham, "Gene Pioneer Urges Dream of Human Perfection," *Globe and Mail* (Toronto), October 27, 2002.

[51] *Human Cloning and Human Dignity,* p. 123.

[52] Ibid., pp. 123–24.

[53] Lee M. Silver, *Remaking Eden: Cloning and Beyond in a Brave New World* (New York: Avon Books, 1997), p. 169.

[54] Ibid., pp. 126–27.

[55] Joseph Fletcher, *The Ethics of Genetic Control: Ending Reproductive Roulette* (Buffalo, New York: Prometheus Books, 1988), p. 177.

Chapter 4

[1] See "Timeline," Biotechnology Industry Organization website, BIO.org.

[2] Carla Dennis, "Chinese Fusion Method Promises Fresh Route to Human Stem Cells," *Nature,* vol. 424 (August 2002), p. 711.

[3] Gina Kolata, "Scientists Create Human Embryos through Cloning," *New York Times,* February 11, 2004.

[4] Correspondence from Carl B. Feldbaum, President of BIO, to Senator Orrin Hatch, dated June 14, 2002, as posted on BIO's website at http://BIO.org/bioethics/moratorium.asp.

[5] S. 303, Sec. 101 (a) (1). The other bill was from Texas, 2003 Session SB 1034.

[6] Remarks by Senator Dianne Feinstein on the floor of the United States Senate, *Congressional Record,* June 14, 2002, p. S5580.

[7] To get around the public's queasiness with human cloning, some researchers have proposed parthenogenesis on unfertilized human oocytes to determine if the procedure can stimulate the development of stem cells, as has occurred in some animal models. If Feintein's bill had been about this type of procedure, her presentation would have had some semblance of accuracy. But the subject of the bill was SCNT. Hence, her description was both unscientific and factually wrong.

[8] Isaac Rabino, "Stem Cell and Cloning Controversies," *Genetic Engineering News,* vol. 24, no. 4 (February 15, 2004), pp. 6–9.

[9] "Genetic Engineering News Reports Results of Stem Cell and Cloning Survey of Scientists," *Business Wire,* February 17, 2004.

[10] Rabino, "Stem Cell and Cloning Controversies," p. 7.

[11] Ibid., p. 9.

[12] The Ethics Committee of the American Society for Reproductive Medicine, "Human Somatic Cell Nuclear Transfer (Cloning)," *Fertility and Sterility,* vol. 74, no. 5 (November 2000), p. 873.

13 Kim Tae-gyu, "Commercial Application of Embryo Cloning Due within Ten Years; Hwang," *Korea Times,* February 19, 2004.

14 Panayiotis M. Zavos, "Human Reproductive Cloning: The Time Is Near," *Reproductive Biomedicine Online,* vol. 6, no. 4 (June 2003).

15 National Bioethics Advisory Commission, *Cloning Human Beings* (Rockville, Maryland, 1997), p. 93 of online edition.

16 The Ethics Committee of the American Society for Reproductive Medicine, "Human Somatic Cell Nuclear Transfer (Cloning)," p. 875.

17 See NAS, *Scientific and Medical Aspects of Reproductive Cloning* (Washington, D.C.: National Academies Press, 2002).

18 Ibid., p. 94.

19 Ibid., p. 97.

20 See, for example, John A. Robertson's *Children of Choice* (Princeton, New Jersey: Princeton University Press, 1994). Robertson is a law professor and bioethicist at the University of Texas, Austin. In his book, the author does not include cloning as part of the right to reproduce. However, he has since expanded his approach to include a right to cloning in some cases.

21 Carson Strong, "Cloning and Infertility," in *The Human Cloning Debate,* ed. Glenn McGee (Berkeley, California: Berkeley Hills Books, 2000), pp. 184– 211; Timothy F. Murphy, "Entitlement to Cloning," in ibid., pp. 212–20.

22 Glenn McGee and Ian Wilmut, "A Model for Regulating Cloning," in *The Human Cloning Debate,* ed. McGee, pp. 221–33.

23 "Dolly's Creator Backs Cloned Babies," *BioEdge,* February 27, 2004.

24 Richard Cohen, "Unsettling, Maybe, But Not Unethical," *Washington Post,* January 2, 2003.

25 Robert P. Lanza, Arthur L. Caplan, Lee M. Silver, Jose B. Cibelli, Michael D. West, Ronald M. Green, "The Ethical Validity of Using Nuclear Transfer in Human Transplantation," *Journal of the American Medical Association,* vol. 284 (December 27, 2000), pp. 3175–79.

26 Marcia Coyle, "The Clone Zone," *National Law Journal,* May 15, 2002.

27 Biotechnology Industry Organization, "The Value of Therapeutic Cloning," © 2003, as found on BIO website, September 8, 2003.

28 For example, see Ben Hirschler, "Stem Cells May Eliminate Need for Heart Transplants," Reuters, September 1, 2003, reporting that in early human trials, four out of five heart transplant candidates treated with their own bone marrow ASCs no longer needed transplant surgery.

29 NAS, "Potential U.S. Populations for Stem-Cell-Based Therapies," copied by author from the NAS website, February 19, 2002.

30 Rick Weiss and Justin Gillis, "New Embryonic Stem Cells Made Available," *Washington Post,* March 4, 2004.

31 David A. Prentice, *Stem Cells and Cloning* (San Francisco: Benjamin Cummings, 2002), p. 26.

32 Peter Aldous, "Can They Rebuild Us?" *Nature,* vol. 410 (April 5, 2001), pp. 622–25.

33 J. S. Odorixo, D. S. Kaufnan, J. A. Thomson, "Multlineage Differentiation from Human Embryonic Stem Cell Lines," *Stem Cells,* vol. 19 (2001), pp. 193–204.

34 Peter Mombaerts, "Therapeutic Cloning in the Mouse," in *Proceedings of the National Academy of Sciences: Regenerative Medicine,* papers from the Arthur M. Sackler Colloquium, held October 18–22, 2002, at the Arnold and Mabel Beckman Center of the National Academy of Sciences and Engineering in Irvine, California (Washington, D.C.: National Academies Press, 2003), pp. 11924–25.

35 Ibid., p. 11925.

36 For example, see "Misguided Chromosomes Foil Primate Cloning," *Science,* vol. 300 (April 11, 2003), pp. 225–27.

37 W. S. Hwang et al., "Evidence of a Pluripotent Human Embryonic Stem Cell Line Derived from a Cloned Blastocyst," *Science,* vol. 303 (March 12, 2004).

38 Mombaerts, "Therapeutic Cloning in the Mouse."

39 Aldous, "Can They Rebuild Us?" (My emphasis.)

40 Ruth Padawer, "Soaring Egg Donation Prices Causing Ethical Concerns," NorthJersey.com, October 8, 2002.

41 BBC News, "Indians Selling Human Organs," October 15, 2002.

42 See Tom Hundley, "Sale of Human Organs Flourishes in Turkey," Knight Ridder News, April 3, 2001.

43 Bernadine Healy, M.D., "The High Cost of Eggs," *U.S. News and World Report,* January 13, 2003.

44 Jo Revill, "IVF Egg Donors 'Risking Their Health,'" *The Observer* (U.K.) February 9, 2003.

45 Carina Dennis, "Chinese Fusion Method Promises Fresh Route to Human Stem Cells," *Nature,* vol. 24, no. 711 (August 14, 2003).

46 Mombaerts, "Therapeutic Cloning in the Mouse."

47 David A. Prentice, interview with author, August 22, 2003.

48 Martin Hutchinson, "Aborted Fetus Could Provide Eggs," BBC News, July 3, 2002.

49 *Cloning Californians? Report of the California Advisory Committee on Human Cloning* (Sacramento, California, 2002), p. 1.

50 Neuroethics was defined at the conference as "the study of the ethical, legal, and social questions that arise when scientific findings about the brain are carried into medical practice, legal interpretations, and health

and social policy." See *Neuroethics: Mapping the Field: Conference Proceedings* (New York: The Dana Foundation, 2002), May 13–14, 2002, p. iii.

51   Ibid., pp. 92–94. (Emphasis added.)

52   For example, see Peter Singer, *Practical Ethics,* 2nd ed. (Cambridge, UK: Cambridge University Press, 1993).

53   State of New Jersey, 210th Legislature, S. 1909, as amended by the Senate Health, Human Services and Senior Citizens Committee on November 25, 2002.

54   Ibid., Section 2 a. (1).

55   Ibid., Section 2 c. (3) (my emphasis).

56   Correspondence dated January 27, 2003 to the Honorable James E. McGreevey, as published in *National Review Online* at www.national review.com, February 3, 2002.

57   78 (r) SB 1034—Introduced version—Bill Text, as copied from the Texas Legislature website on March 18, 2003.

58   Ibid., Subchapter B (a).

59   Senate Bill No. 55, the "Cloning Prohibition and Research Protection Act," Delaware State Senate, 142nd General Assembly.

60   Maryland House Bill 482, Subtitle A.

61   "Cloned Cells Make Kidneys in Cattle, Company Says," Reuters, June 2, 2002.

62   R. Lanza et al., "Generation of Histocompatible Tissue Using Nuclear Transplantation," *Nature Biotechnology,* vol. 20 (July 2002), pp. 689–96. (My emphasis.)

63   *Neuroethics: Mapping the Field,* p. 93.

64   David A. Prentice, interview with author, May 16, 2003.

65   For example, Princeton University bioethicist Peter Singer has suggested that it would have been better to use comatose people rather than chimpanzees in the development of a hepatitis vaccine. See Jill Neimark, "Living and Dying with Peter Singer," *Psychology Today,* January/February 1999, p. 58.

66   "The California Stem Cell Research and Cures Initiative."

67   National Academy of Sciences, "The Value of Therapeutic Cloning for Patients."

68   For example, see Associated Press, "Study: Stem Cell Cloning Flawed," July 5, 2001. "Stem Cells Harbor Genetic Abnormalities, Scientists Find," *Minneapolis Star Tribune,* July 6, 2001. There appears to be significant disagreement on this point in the research community. For example, Rudolf Jaenisch, of the Whitehead Institute, believes that it is "unlikely that genetic instability" of cloned mammals "would block the curative use of embryonic stem cells."

69 Zenit News Agency, "Pushing for a Worldwide Ban on Human Cloning," September 28, 2002, describing the debate in the United Nations over a proposed world treaty that would outlaw cloning of human life.

Chapter 5

1 Tom Abate, interview with author, January 28, 2003.
2 Arthur L. Caplan, "Attack of the Anti-Cloners," *Nation,* June 17, 2002.
3 Glenn McGee, "The Wisdom of Leon the Professional [Ethicist]," *American Journal of Bioethics,* vol. 3, no. 3 (Summer 2003), p. vii.
4 Leon R. Kass, *Life, Liberty and the Defense of Dignity: The Challenge for Bioethics* (San Francisco: Encounter Books, 2002), p. 30.
5 Leon Kass, interview with author, November 6, 2002.
6 McGee, "The Wisdom of Kass," p. viii.
7 Rick Weiss, "Clone Study Casts Doubt on Stem Cells," *Washington Post,* July 6, 2001.
8 Editorial, "Facts versus Ideology in the Cloning Debate," *The Lancet,* vol. 363 (February 21, 2004), p. 581.
9 Woo Suk Hwang et al., "Evidence of a Pluripotent Human Embryonic Stem Cell Line Derived from a Cloned Blastocyst." *Science,* vol. 303 (March 12, 2004), pp. 1669–1674.
10 Australian Broadcasting Corporation, "Korean Stem Cell Research Labeled Recipe for Human Cloning," February 13, 2004.
11 "Facts versus Ideology."
12 Jeffrey M. Drazen, "Legislative Myopia on Stem Cells," *New England Journal of Medicine,* vol. 349, no. 300 (July 17, 2003).
13 Raja Mishra, "Influential Journal Plans Push to Publish More Stem Cell Studies," *Boston Globe,* July 17, 2003.
14 Drazen, "Legislative Myopia."
15 Roger Pielke Jr., "Science Policy: Policy, Politics and Perspective," *Nature,* vol. 416 (March 28, 2002), pp. 367–68.
16 Dennis Shanahan, "Stem-Cell Rat Trick Angers MPs," *The Australian,* August 27, 2002.
17 Sid Marris, "Crippled Rat 'Not Cured,'" *The Australian,* September 2, 2002.
18 Testimony of Peter Rabins, Johns Hopkins University, and John Morris, Washington University, before the Senate Health, Education, Labor and Pensions Subcommittee, May 11, 2004.
19 Rick Weiss, "Stem Cells an Unlikely Therapy for Alzheimer's," *Washington Post,* June 10, 2004.
20 Ibid.
21 Ibid.

22 Stephen M. Barr, "Retelling the Story of Science," *First Things,* March
    2003. See also the author's fine *Modern Physics and Ancient Faith* (Notre
    Dame, Indiana: University of Notre Dame Press, 2003).

23 Francis Fukuyama, *Our Posthuman Future: Consequences of the Biotech-
    nology Revolution* (New York: Farrar, Straus and Giroux, 2002), p. 218.

24 Quoted in Neil Munro, "Science Policy—Securing Science," *National
    Journal,* September 6, 2003.

25 Source: www.meta-library.net.

26 Barr, "Retelling the Story of Science."

27 Stuart Newman, interview with author, September 25, 2002.

28 Iain Murray, "The Return of Scientism," Tech Central Station, September
    6, 2002, quoting Baroness Mary Warnock of Cambridge University, the
    "founding mother" of the United Kingdom's Human Fertilization and
    Embryology Authority, after it authorized embryo screening to ensure
    that a baby would be born that could provide tissues to help treat an ill
    sibling.

29 Rahul K. Dhanda, *Guiding Icarus: Merging Bioethics with Corporate Inter-
    ests* (New York: Wiley, 2002), p. 234.

30 Michael Shermer, "The Shamans of Scientism," *Scientific American,* June
    2002, p. 35.

31 For example, see Carmel Shalev, "Reproductive Cloning—A Human
    Rights Framework," presented to Ad Hoc Committee on an International
    Convention Against the Reproductive Cloning of Human Beings, Febru-
    ary 25, 2002.

32 Paul Copland, "Science and Ethics Must Not Be Separated: The Progress
    of Research Must Be Kept Free from Religious and Political Interven-
    tion," *Nature,* vol. 425 (September 11, 2003), p. 121.

33 Munro, "Science Policy—Securing Science."

34 Neil Munro, "Petri Dish Politics," *National Journal,* April 19, 2003.

35 Stephen S. Hall, "Eve Redux: The Public Confusion over Cloning,"
    Hastings Center Report, May/June 2003.

36 Nebraskans for Research, "Who Are David A. Prentice and Wesley
    Smith?" Advocacy memorandum passed out to Nebraska legislators after
    we both testified before a legislative committee on January 25, 2002, in
    favor of a state ban on human cloning.

37 David A. Prentice, interview with author, January 5, 2004.

38 Ibid.

39 Mark Dowie, "Biotech Critics at Risk," *San Francisco Chronicle,* January
    11, 2004.

40 Quoted in Munro, "Petri Dish Politics."

41 Edwin Black, *War Against the Weak: Eugenics and America's Campaign to*

Create a Master Race (New York: Four Walls Eight Windows, 2003),
p. 107. This may be the best book on the American eugenics movement.

42 Ibid., p. ix.

43 Ibid., p. 58.

44 Daniel V. Kelves, *In the Name of Eugenics: Genetics and the Uses of Human
Heredity* (Cambridge, Massachusetts: Harvard University Press, 1985),
p. 10.

45 Ibid., p. 89.

46 Diane B. Paul, *Controlling Human Heredity, 1865 to the Present* (Atlantic
Highlands: Humanities Press International, 1995), p. 11.

47 *Buck v. Bell*, 274 U.S. 200 (1927).

48 Black, *War Against the Weak*, p. 123.

49 Ibid., p. 247.

50 For an excellent account of the early euthanasia movement's close ties to
the eugenics movement, see Ian Dowbiggin, *A Merciful End: The
Euthanasia Movement in Modern America* (New York: Oxford University
Press, 2003).

51 Black, *War Against the Weak*, p. 443.

52 Ibid., p. 441.

53 Sheldon Krimsky, "Science on Trial: Conflicts of Interest Jeopardize
Scientific Integrity and Public Health," *Gene Watch*, vol. 3, no. 5
(September/October 2003).

54 Neil Munro, "An Interested Disinterest in Cloning," *National Journal*,
September 8, 2001.

55 Neil Munro, "Doctor Who?" *Washington Monthly*, November 2002.

56 Neil Munro, "Mixing Business with Stem Cells," *National Journal*, July
21, 2001.

57 Fred L. Bookstein, "Biotech and the Watchdog Role of Universities,"
Washington Post, July 30, 2001.

58 Munro, "Doctor Who?"

59 Jordan Mejias, "Research Always Runs the Risk of Getting Out of Control," *Frankfurter Allgemeine Zeitung*, June 4, 2001. Chargaff's quip was
taken from the American Philosophical Society, "Erwin Chargaff Papers,"
Philadelphia.

60 David P. Hamilton, "Biotech's Dismal Bottom Line: More Than $40 Bil-
lion in Losses," *Wall Street Journal*, May 20, 2004.

61 "Biotech Industry Has Large Cash Reserves," Xinhua News Agency, June
10, 2002.

62 Hamilton, "Biotech's Dismal Bottom Line."

63 Philip R. Reilly, Foreword to Rahul K. Dhanda, *Guiding Icarus: Merging
Bioethics with Corporate Interests* (New York: Wiley, 2002), p. xi.

[64] Robert A. Weinberg, "Of Clones and Clowns," *Atlantic Monthly,* June 2002, p. 58.

[65] Rick Weiss, "For Me, It Was Never about God," *Oakland Tribune,* November 24, 2002. In "Bush Unveils Bioethics Council: Human Cloning," *Washington Post,* January 17, 2002, Rick Weiss wrote: "In November, researchers announced that they had made the first human embryo clones, giving immediacy to warnings by religious conservatives and others that science is no longer serving the nation's moral will. At the same time, the United States was fighting a war to *free a faraway nation from the grip of religious conservatives who were denounced for imposing their moral code on others*. . . . Until now, opponents of therapeutic cloning have largely made their case on the grounds that it would be difficult to stop someone from making a cloned baby if it were legal to make cloned embryos. But experts said that if the new panel supports the other major line of reasoning—that human embryos are inherently deserving of protections—such support could legitimate an effort to *codify fundamentalist views* into law." (My emphasis.)

[66] For example, see Gareth Cook, "Patent Hints How Firm May Farm Human Tissue," *Boston Globe,* July 26, 2001, and Marilynn Marchione, "Ethical, Legal Questions Hardly Sway Scientist at Vanguard of Human Cloning," *Milwaukee Journal Sentinel,* May 5, 2002. Also, Center for Genetics and Society, "The Biotechnology Industry," analysis of ACT self-promotion, at http://www.genetics-and-society.org/analysis/biotech.html.

[67] Joannie Fischer, "The First Clone," *U.S. News and World Report,* December 3, 2001.

[68] Ibid.

[69] For example, Ronald M. Green, *The Human Embryo Research Debates* (New York: Oxford University Press, 2001), writes that "political interference" and "obstruction"—by which he means refusal to federally fund—with regard to embryonic stem cell research "must end."

[70] Ibid., pp. 62–63.

[71] Arlene Weintraub and Faith Keenan, "The Clone Wars," *Business Week,* March 15, 2002.

[72] Gary Stix, "What Clones?" *Scientific American,* December 31, 2001.

[73] Ibid.

[74] Rudolf Jaenisch, "To Clone or Not to Clone," *Scientific American,* letter to editor, May 2002.

[75] Kyla Dunn, "Cloning Trevor," *Atlantic Monthly,* June 2002.

[76] Ibid., p. 46.

77 Wendy Goldman Rohm, "Seven Days of Creation: The Inside Story of a Human Cloning Experiment," *Wired,* January 2004.

78 "Cloning and Stem Cells: Baby Steps," *The Economist,* December 30, 2003, p. 53. Cibelli is reportedly researching on parthenogenesis, which would seek to extract embryonic stem cells from unfertilized eggs stimulated into dividing without fertilization or SCNT.

79 Rohm, "Seven Days of Creation."

80 PR News Release, "Boston Researchers Create Human Clone Embryo for Therapuetic Use, *Wired* Magazine Reports," December 16, 2003.

81 Rohm, "Seven Days of Creation."

82 Ruth Shalit, "When We Were Philosopher Kings," *New Republic,* April 28, 1997.

83 Carl Elliot, "Throwing the Watchdog a Bone," *Hastings Center Report,* vol. 31, no. 2 (March/April 2003), p. 10.

84 Daniel Callahan, "Doing Good and Doing Well," *Hastings Center Report,* vol. 31, no. 2 (March/April 2003), p. 20.

85 Ibid., pp. 20–21.

86 Vicki Brower, "Biotechs Embrace Bioethics," www.BioSpace.com, June 14, 1999.

87 Philip R. Reilly, Foreword to Dhanda, *Guiding Icarus,* p. xi.

88 Ibid., p. 7.

89 Ibid., pp. 216–17.

90 Andrea Knox, "Ethicist Spurs Debate on Biological Research," *Philadelphia Inquirer,* July 17, 2001.

91 Dhanda, *Guiding Icarus,* p. 227.

92 William Saletan, "The Ethicist's New Clothes," *Slate,* August 17, 2001.

93 David P. Hamilton, "Biotech's Dismal Bottom Line," *Wall Street Journal,* May 20, 2004.

94 "Biotech Industry Has Large Cash Reserves," Xinhua News Agency, June 10, 2002.

95 Penni Crabtree, "Bioethics Getting into Politics Now," *San Diego Union Tribune,* August 12, 2003.

96 Hamilton, "Biotech's Dismal Bottom Line."

97 Luke Timmerman, "Stem-Cell Research Is Exciting, but Not to Investors," *Miami Herald,* March 30, 2004, reprinting an article that originally appeared in the *Seattle Times.*

98 David Firn and Victoria Griffith, "Inside Track: Stem Cell Science on a Shoe String," *Financial Times,* July 18, 2002.

99 Andrew Pollack, "Ethics Aside, a Good Business Model Remains Elusive for Stem Cells," *New York Times,* July 28, 2001. (My emphasis.)

100 Dunn, "Cloning Trevor," p. 43.

[101] James Burns, president of Osiris Therapeutics, Inc., quoted in Richard Miniter, "Hard Cell: Science Doesn't Need Subsidized Embryo Research," OpinionJournal.com (from the *Wall Street Journal*), July 23, 2001.

[102] Timmerman, "Stem-Cell Research Is Exciting, but Not to Investors."

[103] Lydia Saad, "'Cloning' Humans Is a Turnoff to Most Americans: Embryonic Cloning for Research Is Also Opposed," *Gallup News Service*, May 16, 2002.

[104] For example, see discussion of Hatch/Feinstein bill in Chapter 4.

[105] Tracey L. Regan, "Funding Key in Stem Cell Research Fight," *Trenton Times*, January 12, 2004.

[106] "The California Stem Cell Research and Cures Initiative." For more details, see Wesley J. Smith, "California Cloning May Be Coming Soon," *San Francisco Chronicle*, February 1, 2004.

[107] Correspondence from Elizabeth G. Hill, California legislative analyst to California Attorney General Bill Lockyer, December 12, 2003.

Chapter 6

[1] J. Bottum, "The Pig-Man Cometh," *Weekly Standard*, October 23, 2000.

[2] Ibid.

[3] The bioethics historian Albert R. Jonsen first called Fletcher the "patriarch of bioethics" in: "'O Brave New World': Rationality in Reproduction," in *Birth to Death: Science and Bioethics*, ed. David C. Thomasma and Thomasine Kushner (Cambridge, UK: Cambridge University Press, 1996), p. 50.

[4] Joseph Fletcher, *The Ethics of Genetic Control: Ending Reproductive Roulette* (Buffalo, New York: Prometheus Books, 1988), pp. 3–4.

[5] Ibid., p. 45.

[6] Philip Kitcher, *The Lives to Come: The Genetic Revolution and Human Possibilities* (New York: Touchstone, 1997), p. 202.

[7] Carolyn Abraham, "Gene Pioneer Urges Dream of Human Perfection," *Globe and Mail* (Toronto), October 26, 2002.

[8] Tony Blankley, "Cloning and the Chinese," TownHall.com, September 24, 2003.

[9] Gregory E. Pence, *Who's Afraid of Human Cloning?* (New York: Rowman & Littlefield Publishers, 1998), pp. 101–2.

[10] Gregory Stock, *Redesigning Humans: Our Inevitable Genetic Future* (New York: Houghton Mifflin Company, 2002).

[11] Ibid., p. 185.

[12] Lee M. Silver, *Remaking Eden: Cloning and Beyond in a Brave New World* (New York: Avon Books, 1997), p. 129.

[13] Ibid., p. 237.

14   Ibid., p. 6.

15   Ibid., pp. 249–50.

16   Nick Bostrom, "The Transhumanist FAQ," World Transhumanist Association.

17   World Transhumanist Association, "About the WTA," http://transhumanism.org.

18   James J. Hughes, "The Future of Death: Cryonics, and the Telos of Liberal Individualsm," *Journal of Evolution and Technology*, vol. 6 (July 2001).

19   Simon Smith, "Transcending Humanism," Betterhumans.com, January 26, 2003 (quoting Patrick Inniss).

20   Leigh Turner, "Biotechnology as Religion," *Nature Biotechnology*, vol. 22 (2004), pp. 659–60.

21   Hughes, "The Future of Death."

22   World Transhumanist Association, "The Transhumanist Declaration," obtained at www.transhumanism.org.

23   Erik Baard, "Cyborg Liberation Front," *Village Voice*, July 30/August 5, 2003.

24   Hughes, "The Future of Death."

25   Ibid.

26   James J. Hughes, "Democratic Transhumanism," obtained at www.changesurfer.com/Acad/DemocraticTranshumanism.htm.

27   Leon R. Kass, "The Moral Meaning of Genetic Technology," *Commentary*, September 1999, p. 38.

28   Jason Scott Robert and Françoise Baylis, "Crossing Species Boundaries," *American Journal of Bioethics*, vol. 3, no. 3 (Summer 2003), p. 4.

29   Ibid., p. 6.

30   A. M. Chakrabarty, "Crossing Species Boundaries and Making Human-Nonhuman Hybrids: Moral and Legal Ramifications," *American Journal of Bioethics*, pp. 20–21.

31   Mihail C. Roco and William Sims Bainbridge, eds., *Converging Technologies for Improving Human Performance* (Arlington, Virginia: National Science Foundation, 2002).

32   Mihail C. Roco and William Sims Bainbridge, "Overview: Converging Technologies for Improving Human Performance," in ibid., p. 22.

33   Ibid., p. 19.

34   Catherine Orenstein, "Stepford Is Us," *New York Times*, June 9, 2004. *The Stepford Wives* is a clone of that clone classic movie of the same name.

35   Lisa M. Krieger, "Ethical Issues in Mixing Brain Cells," *San Jose Mercury News*, December 8, 2002.

36 Nicholas D. Kristof, "Interview with a Humanoid," *New York Times,* July 23, 2002.

37 Associated Press, "Pigs Created That Carry Human Genes," as published in *New York Times,* October 22, 2002.

38 National Academy of Sciences, Board of Life Sciences, *Animal Biotechnology: Identifying Science-Based Concerns* (Washington, D.C.: National Academies Press, 2002).

39 Ian Wilmut, Keith Campbell and Colin Tudge, *The Second Creation: Dolly and the Age of Biological Control* (Cambridge, Massachusetts: Harvard University Press, 2000), p. 252.

40 The President's Council on Bioethics, *Beyond Therapy: Biotechnology and the Pursuit of Happiness* (Washington, D.C.: President's Council on Bioethics, 2003), pp. 33–34.

41 Rifkin, *Biotech Century* (New York: Jeremy P. Tarcher/Putnam, 1999), p. 140, citing a poll published in *Newsweek,* Special Issue, "The Twenty-First Century Family," Spring 1990, p. 98.

42 President's Council, *Beyond Therapy,* p. 34.

43 Ibid., p. 37.

44 Ibid., p. 38.

45 Ibid.

46 I have taken the term "screening-out" from the President's Council, *Beyond Therapy.*

47 Oliver Morton, "Overcoming Yuk," *Wired,* January 1998.

48 Ruth Hubbard and Stuart Newman, "Yuppie Eugenics," *Z Magazine,* March 2002.

49 Reuters, "British Couple Get Green Light for 'Designer' Baby," February 22, 2002, reporting on a British couple's plan to use genetic embryo screening to select an embryo for implantation whose umbilical-cord blood could be used to treat their existing son's genetic disease.

50 Lindsey Arent, "Serving Up Eggs on the Web," *Wired,* August 26, 1999.

51 Associated Press, "Sperm Bank for the Elite Reports Its First Birth," as published in the *Boston Globe,* May 24, 1982.

52 Stock, *Redesigning Humans,* p. 14.

53 Jere Longman, "Getting the Athletic Edge May Mean Altering Genes," *New York Times,* May 11, 2001.

54 For more on this matter, see Wesley J. Smith, *Culture of Death: The Assault on Medical Ethics in America* (San Francisco: Encounter Books, 2001).

55 Pence, *Who's Afraid of Human Cloning?* pp. 167–68.

56 See for example, Allen Buchanan, Dan W. Brock, Norman Daniels and

Daniel Wikler, *From Chance to Choice: Genetics and Justice* (Cambridge, UK: Cambridge University Press, 2002).

57 Philip Kitcher, *The Lives to Come,* pp. 202–4.

58 John A. Robinson, *Children of Choice: Freedom and the New Reproductive Technologies* (Princeton, New Jersey: Princeton University Press, 1994), p. 150.

59 Ibid., pp. 150–51.

60 Ibid., p. 170.

61 Ibid., p. 153.

62 See Peter Singer, *Rethinking Life and Death: The Collapse of Our Traditional Ethics* (New York: St. Martin's Press, 1994).

63 Jean E. Chambers, "Women's Right to Choose Rationally: Genetic Information, Embryo Selection, and Genetic Manipulation," *Cambridge Quarterly of Healthcare Ethics,* vol. 12 (2003), p. 420.

64 Ibid.

65 Ibid., p. 426.

66 "Turning Point for the Human Species," interview with George Annas, *Trial,* July 2001, p. 24.

67 Eric Cohen, "New Genetics, Old Quandaries," *Weekly Standard,* April 22, 2002, p. 26.

68 Bill McKibben, *Enough: Staying Human in an Engineered Age* (New York: Henry Holt and Company, 2003), p. 33.

69 Science and Religion Information Services, "European Scholars Support Development of Germ Line Modification," December 13, 2002.

70 As quoted in "DNA Discoverer: Use Genetics to Improve IQ," Agence France-Presse, February 28, 2003.

71 Abraham, "Gene Pioneer Urges Human Perfection."

72 McKibben, *Enough: Staying Human,* p. 35.

73 Ibid., p. 59.

74 Ibid., p. 58.

75 Ibid., p. 59.

76 Fletcher, *The Ethics of Genetic Control,* p. 157.

77 Joseph Fletcher, *Humanhood: Essays in Biomedical Ethics* (Buffalo, New York: Prometheus Books, 1979), p. 118.

78 Ibid., p. 119.

79 Fletcher, *The Ethics of Genetic Control,* p. 61.

80 Edwin Black, *War Against the Weak: Eugenics and America's Campaign to Create a Master Race* (New York: Four Walls Eight Windows, 2003), p. 429.

81 Robert Williamson and Rony Duncan, "DNA Testing for All," *Nature,* vol. 418 (2002), pp. 585–86.

82 Lois Rogers, "Having Disabled Babies Will Be a 'Sin,' Says Scientist," *Sunday Times* (London), July 4, 1999.

83 Black, *War Against the Weak*, p, 431.

84 Phil Bereano and Richard Sclove, "Life, Liberty and the Pursuit of Genetic Testing," *Washington Post*, March 22, 1998.

85 Julian Savulescu, "Should Doctors Intentionally Do Less Than the Best?" *Journal of Medical Ethics*, vol. 25 (1999), pp. 121–26.

86 Ibid., p. 125.

87 Ibid., p. 126.

88 Fletcher, *Humanhood*, p. 118.

89 Ibid., pp. 123–24. (My emphasis.)

90 Fletcher, *The Ethics of Genetic Control*, pp. 3, 28.

91 Ibid., pp. 4, 200.

92 Pence, *Who's Afraid of Human Cloning?* p. 175.

93 Ibid., pp. 101–2.

94 Ibid., p. 114.

95 Victoria Button, "Control Gene Pool, Says Ethicist," *The Age*, October 13, 2000.

96 Buchanan et al., *From Chance to Choice*, p. 336.

97 Ibid., pp. 336–37, 338. It is worth noting that the People's Republic of China already engages in compulsory eugenics. See Frank Dikotter, *Imperfect Conceptions: Medical Knowledge, Birth Defects, and Eugenics in China* (New York: Columbia University Press, 1998).

98 Baard, "Cyborg Liberation Front."

99 Ibid.

100 As quoted in William Kristol and Eric Cohen, eds., *The Future Is Now: America Confronts the New Eugenics* (Lanham, Maryland: Rowman & Littlefield Publishers), p. 46. Excerpts from Paul Ramsey, *Fabricated Man: The Ethics of Genetic Control* (New Haven: Yale University Press, 1970).

Chapter 7

1 Leon Kass has evocatively written on the philosophical concept of human dignity. See, for example, Leon R. Kass, *Life, Liberty and the Defense of Dignity: The Challenge for Bioethics* (San Francisco: Encounter Books, 2002).

2 Leon R. Kass, "Preventing a Brave New World," *New Republic*, May 21, 2001.

3 Michael Fumento, *BioEvolution: How Biotechnology Is Changing Our World* (San Francisco: Encounter Books, 2003).

4 Amy Fagan, "Adult Stem Cells Produce Treatment Breakthroughs,"

Washington Times, December 29, 2003, quoting a study published in the May 2003 *Nature Medicine.*

5   Martha S. Windrem et al., "Fetal and Adult Human Oligodendrocyte Progenitor Cell Isolates Myelin in the Congenitally Dysmyelinated Brain," *Nature Medicine,* vol. 10, no. 1, pp. 93–97. The study also found that late-stage fetal cells helped produce myelination, although not as efficiently as the adult stem cells.

6   "Muscular Dystrophy: Blood Cells Could Build Muscle in Neuromuscular Diseases," *Health and Medicine Week,* December 1, 2003.

7   Oliver Wright, "Patient's Own Skin Cells Turned into Potential Alzheimer's Treatment," *The Times* (London), December 10, 2003.

8   American Society of Hematology, "Derivation of Functional Insulin-Producing Cells from Human Bone Marrow-Derived Stem Cells," press release, December 8, 2003.

9   Shota Kodama et al., "Islet Regeneration during Reversal of Autoimmune Diabetes in NOD Mice," *Science,* vol. 302 (November 14, 2003), p. 1223.

10 David Stonehouse, "Not Just Blind Hope: Stem Cell Therapy Sets Sight on a Cure," *Edmonton Journal,* November 30, 2003.

11 Gabe Romain, "Fat's Full of True Stem Cells," Betterhumans.com, March 8, 2004.

12 Correspondence from James F. Battey Jr., Chairman, NIH Stem Cell Task Force, to Representatives Mark Souder and Christopher Smith, June 18, 2004.

13 Helen R. Pilcher, "GM Skin Cells Help Treat Alzheimer's?" *Nature Science Update,* April 28, 2004.

14 "Dental Pulp May Hold Key to Treatment of Parkinson's," University of Michigan News Service, May 4, 2004.

15 Yoon-Young Jan et al., "Hematopoietic Stem Cells Convert into Liver Cells within Days without Fusion," *Nature Cell Biology,* June 1, 2004.

16 Franco Locatelli et al., "Related Umbilical Cord Blood Transplant in Patients with Thalassemia and Sickle Cell Disease," *Blood,* November 7, 2002.

17 Susan L. Staba et al., "Cord-Blood Transplants from Unrelated Donors in Patients with Hurler's Syndrome," *New England Journal of Medicine,* vol. 350, no. 19 (May 6, 2004).

18 Svitlana Garbuzova-Davis et al., "Intravenous Administration of Human Umbilical Cord Blood in a Mouse Model of Amyotrophic Lateral Sclerosis: Distribution, Migration, and Differentiation," *Journal of Hematotherapy and Stem Cell Research,* vol. 12 (2003), pp. 255–70.

19 Fumento, *BioEvolution,* pp. 256–70.

20 "Rats with Partial Parkinson's Damage in the Brain Show Complete

Recovery after Gene Therapy," *University of Florida News*, March 26, 2002.

21 Malcolm Ritter, "Successful Gene Therapy Reported on Cells Key to Hearing Loss," Associated Press, as published in the *San Francisco Chronicle*, June 1, 2003.

22 William B. Hurlbut, "Statement of Dr. Hurlbut," *Human Cloning and Human Dignity: The Report of the President's Council on Bioethics* (New York: Public Affairs, 2002), pp. 310–20.

23 Timothy Noah, "Leon Kass, You Silly Ass!" *Slate*, March 8, 2004.

24 Charles Krauthammer, "Crossing Lines," *New Republic*, April 29, 2002, pp. 21, 23.

25 Gregory Stock, *Redesigning Humans: Our Inevitable Genetic Future* (New York: Houghton Mifflin Company, 2002), p. 2.

26 Leon R. Kass, "Ageless Bodies, Happy Souls," *New Atlantis*, Spring 2003, p. 20.

27 See, for example, Dianne Irving, "Playing God by Manipulating Man: The Facts and Frauds of Human Cloning," presented at Missouri Catholic Conference Workshop, October 4, 2003, as published by LifeIssues.Net.

28 Leila Abboud, "FDA Seeks Rigorous Review of New Fertility Treatments," *Wall Street Journal*, October 7, 2002.

29 Gina Kolata, "Using Genetic Tests, Ashkenazi Jews Vanquish a Disease," *New York Times*, February 18, 2003.

30 The Board of Directors of the Council for Responsible Genetics, "The Genetic Bill of Rights: Right 10," Spring 2000.

31 Neil Munro, "Technology: The New Patent Puzzle," *National Journal*, March 2, 2002.

32 William Kristol, "Brave New Patents," *Weekly Standard*, May 27, 2002.

33 International Center for Technology Assessment, "Patent Watch Discloses Three Pending Patents Which Would Include Cloned Human Embryos and Fetuses," press release, May 16, 2002. (Emphasis in *Patent Watch*.)

34 Reuters, "European Patent Office Bars Human Cloning in Patent," July 24, 2002.

35 Biotechnology Industry Organization, "New Patent Legislation Sets Dangerous Precedent and Stifles Research," *Biotechnology Information, Advocacy and Business Support*, September 2, 2003.

36 "Patenting Pieces of People," *Nature Biotechnology*, vol. 21, no. 4 (April 2003).

37 For example, see "Stealth Stipulation Shadows Stem Cell Research," *Scientist*, vol. 18, no. 4 (March 1, 2004).

38 Apparently nothing is too bizarre to be contemplated by some biotecho-philes. The prospect of implanting human embryos into nonhuman wombs has been addressed by the President's Council on Bioethics. See "Staff Working Paper," Draft Recommendations (Revised), discussed at the council's January 2004 meeting.

39 Kim Lunman, "Senate Passed 'Historic' Bill on Reproductive Technology," *Globe and Mail* (Toronto), March 12, 2004.

40 For example, see Jo Revill, "IVF Donors 'Risking Their Health,'" *The Guardian* (Manchester), February 9, 2003.

41 Ruth Padawer, "Soaring Egg Donation Prices Causing Ethical Concerns," NorthJersey.com, October 8, 2002.

42 Phil Bereano and Richard Sclove, "Life, Liberty, and the Pursuit of Genetic Testing," *Washington Post*, March 22, 1998.

43 Jesse J. Holland, "Advocates Argue over Genetic Privacy," Associated Press, September 13, 2002.

44 David E. Sanger, "Bush Pledges to Restrict DNA Discrimination," *San Francisco Chronicle*, June 24, 2001; from an article that originally appeared in the *New York Times*.

45 Gregory E. Pence, *Who's Afraid of Human Cloning?* (Lanham, Maryland: Rowman & Littlefield Publishers, 1998), p. 170.

46 Robert Pollack, "New Genetics Drain on Public Health," *San Jose Mercury News*, July 8, 2001.

47 *Monitoring Stem Cell Research: A Report of the President's Council on Bioethics*, pre-publication version (Washington, D.C., 2004), p. 129.

Index

McKay, Ronald D. G., 100
McKibben, Bill, 147, 149, 150, 162n
media: on adult stem cells, 116–17;
on embryonic stem cells, 25–26,
39, 40, 43; on human cloning,
61–62, 97, 158–59; and industry
PR, 112–19; scientific inaccuracy
in, 97–99, 113, 117–19
medical treatment: *see* regenerative
medicine
Mombaerts, Peter, 79–80, 82
Moore, Mary Tyler, 26, 27
Morton, Oliver, 142
Mountford, Peter, 128
multiple sclerosis, 42, 159
Munro, Neil, 110
muscular dystrophy, 159

nanotechnology, 131, 133, 137
National Academy of Sciences
(NAS), 7, 56, 77, 78, 79, 101,
103; on cloning, 74, 110
National Bioethics Advisory Com-
mission (NBAC), 9–10; cloning
moratorium, 74; on ESC research
funding, 9, 24–25
National Institute of Neurological
Disorders and Stroke, 100
National Institutes of Health, 24, 25,
106
National Journal, 40, 101, 103, 105,
106, 110, 167
National Law Journal, 76
National Science Foundation, 137
Nature, xiii–xiv, 32, 80, 82, 99, 103,
151
Nature Biotechnology, 134, 168–69
Nazis, 19, 106, 107
Nebraska, 105
Netherlands, 37
Neuhaus, Richard John, 21

Neurology, 12
New Atlantis, 147
New England Journal of Medicine,
97–98
New Jersey: cloning law (S. 1909),
85–86, 88, 89, 124–25, 157, 170
Newman, Stuart, 55, 102, 142
New Republic, 119
Nietzsche, Friedrich, 5
Nightlight Christian Adoptions, 28
Noah, Timothy, 164
nonembryonic stem cells. *See* adult
stem cells
Norway, 52, 91, 171

Observer, The, 2
Okarma, Thomas, 121
ooplasm transfer, 166
Orenstein, Catherine, 137–38
organ harvesting, 38
organ sale, 81
organ transplantation, 81; from
animals, 20, 139, 162
osteoporosis, 42
Our Posthuman Future (Fukuyama),
101

Parkinson's disease, 7–8, 14, 100,
114, 163; ASC treatment, 42, 100,
106, 159, 161
patents, 109–11, 123, 167–69
Pauling, Linus, 151
Pellegrino, Edmund, 26
Pence, Gregory E., 171; on cloning,
58, 37, 131, 144; on reproductive
"rights," 153–54
"personhood" theory, 35–38
pharmaceutical companies, 109, 120,
122
"pharming," 139
Pharming, 122